WAR AND THE ENGINEERS

A volume in the series

CORNELL STUDIES IN SECURITY AFFAIRS

edited by Robert J. Art
　　　　Robert Jervis
　　　　Stephen M. Walt

A full list of titles in the series appears at the end of the book.

WAR AND THE ENGINEERS

The Primacy of Politics over Technology

Keir A. Lieber

CORNELL UNIVERSITY PRESS

Ithaca and London

First published 2005 by Cornell University Press

Printed in the United States of America

Library of Congress Cataloging-in-Publication Data

Lieber, Keir A. (Keir Alexander), 1970–
    War and the engineers : the primacy of politics over technology / Keir A. Lieber.
        p.   cm.—(Cornell studies in security affairs)
    Includes bibliographical references and index.
    ISBN-13: 978-0-8014-4383-1 (cloth : alk. paper)
    ISBN-10: 0-8014-4383-0 (cloth : alk. paper)
    1. Military art and science—Technological innovations—History—19th century—
Case studies.   2. Military art and science—Technological innovations—History—20th
century—Case studies.   3. Military weapons—History—19th century—Case studies.
4. Military weapons—History—20th century—Case studies.   5. Security, International—
History—19th century—Case studies.   6. Security, International—History—20th
century—Case studies.   7. Military art and science—Political aspects—Case studies.
I. Title.   II. Series.
    U41.L54   2005
    355.02—dc22   2005016065

Cornell University Press strives to the use environmentally responsible suppliers and
materials to the fullest extent possible in the publishing of its books. Such materials
include vegetable-based, low-VOC inks and acid-free papers that are recycled, totally
chlorine-free, or partly composed of nonwood fibers. For further information, visit
our website at www.cornellpress.cornell.edu.

Cloth printing   10 9 8 7 6 5 4 3 2 1

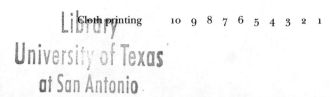

For Meredith

Contents

Acknowledgments

I am indebted to John Mearsheimer for years of patient guidance, constructive criticism, lively debate, support, and friendship. He is inimitable, but also an inspiring role model. My years at the University of Chicago were formative ones, and I am grateful to many teachers and colleagues in the Department of Political Science for fostering a stimulating, nonhierarchical, and unabashedly intellectual environment. I learned a tremendous amount from Stephen Walt, Charles Glaser, and Charles Lipson, all of whom continue to lend unfailing support and valuable guidance at every turn.

This book has deep roots, and many people gave helpful comments over the years. I am grateful for incisive and constructive comments on the entire manuscript provided by Robert Jervis and an anonymous reviewer. Christopher Layne read the manuscript at a late stage and offered helpful comments on theory, history, and strategy. Gerard Alexander, David Edelstein, and Daniel Lindley read individual chapters and gave prompt and insightful advice. For comments on related papers and presentations, I thank David Bearce, Stephen Biddle, Jason Castillo, Timothy Crawford, Alexander Downes, David Edelstein, Bernard Finel, Benjamin Frankel, Michael Freeman, Charles Glaser, Matthew Kocher, Daniel Kryder, Scott Lasensky, Peter Lavoy, Charles Lipson, Sean Lynn-Jones, David McIntyre, John Mearsheimer, Takayuki Nishi, Michael O'Hanlon, Robert Pape, Dan Reiter, Benjamin Riddleberger, Sebastian Rosato, Jordan Seng, Scott

Sagan, Randall Schweller, Stephen Van Evera, Stephen Walt, Alexander Wendt, and Paul Yingling. Charles Glaser, Robert Jervis, Sean Lynn-Jones, and Stephen Van Evera deserve special thanks for aiding and abetting a (partial) critic of their own works. I also thank fellow participants and commentators on my work in the University of Chicago Program on International Politics, Economics, and Security (PIPES) and Program on International Security Policy (PISP).

I owe a special debt of gratitude to Richard Haass and Michael O'Hanlon for a research fellowship in Foreign Policy Studies at the Brookings Institution and to Michael Brown for a visiting faculty position in the Security Studies Program at Georgetown University. Both opportunities came at crucial times and allowed me to complete an initial draft of this work. I also thank Sarah Chilton, Reference Librarian at the Brookings Institution.

For generous financial support of my research and writing, I thank the Brookings Institution, Earhart Foundation, Institute for the Study of World Politics, John D. and Catherine T. MacArthur Foundation, Andrew W. Mellon Foundation, and the Smith Richardson Foundation. I also thank the University of Chicago for a Century Scholarship.

I am indebted to the University of Notre Dame, and many colleagues, administrators, and students, for support and encouragement as I completed the book. I thank the Graduate School Office of Research and the Institute for Scholarship in the Liberal Arts for generous research support, my colleagues in the Department of Political Science for moral support, Cheryl Gray for logistical support, and Ozlem Kayhan and Harsh Pant for research assistance.

At Cornell University Press, I am grateful to Roger Haydon for guiding me through the review and publication process with tact and humor, Jamie Fuller for her superb job at copyediting the manuscript, and Karen Laun for deftly shepherding the manuscript through editing and production. I also thank Carolyn Sherayko for preparing the index. I am fortunate to have worked with such professionals.

My family has been my principal source of support, strength, and inspiration. Joel and Susan Bowers never hesitated about their son-in-law's career or career prospects; for that and much more I am profoundly grateful. My parents, Robert and Nancy Lieber, instilled in me a zest for life and taught me the value of knowledge and intellect. Their optimism and encouragement at every step have made all the difference. I also thank my dad for routinely dropping everything to read and comment on my work. I thank my girls, Sophie, Isabel, and Lucy, for making it so easy to keep everything in perspective. Most important, I thank my wife, Meredith Bowers, for making my life complete and fulfilling.

WAR AND THE ENGINEERS

Introduction

Offense, Defense, and the Prospects for Peace

W
hat is the impact of technology on the politics of war and peace? Do some technological developments provoke war? Do others promote peace? Scholars and other analysts have expended great effort to explain the broader topic of the role of technology in the evolution of human society, as well as the narrower subject of the impact of weapons on warfare,[1] but we have few clear frameworks to make sense of the international political consequences of technological change.

This gap in our knowledge is perhaps unsurprising given the dizzying and unabated pace of technological innovation, which has wrought the industrial, nuclear, and information and communication revolutions in just the last two centuries. Nevertheless, many of the most pressing international security debates today—about the consequences of nuclear weapons proliferation, transnational terrorism and counterterrorism, U.S. power preponderance and national missile defense, and the so-called revolution in military affairs, for example—turn on the likely political effects of emerging technology. Given the enormous stakes and costs entailed in contemporary international conflict, these effects deserve closer scrutiny.

This book explores the relationship between technological change and international relations through an evaluation of "offense-defense theory," the dominant explanation of the topic in political science. Offense-defense theory contends that international conflict and war are more likely to occur when offensive military operations have the advantage over defensive operations, whereas cooperation and peace are more likely when defense dominates. According to the theory, the relative ease of attack and defense—the "offense-defense balance"—is determined primarily by the prevailing state of technology at any given time. When technological change shifts the balance toward offense, attackers are more likely to win quick and decisive victories. Leaders' perceptions of the prospect of quick and decisive warfare exacerbate international security competition and make wars of expansion, prevention, and preemption more likely. On the other hand, when leaders believe technological innovation strengthens the defense, states are more likely to feel secure and act benignly.[2]

Offense-defense theory has had a pervasive and enduring influence on the study of international relations, in large part because it offers an attractively simple, powerful, and intuitively plausible explanation for conflict and cooperation. One scholar has described offense-defense theory as "the most powerful and useful realist theory on the causes of war."[3] For example, offense-defense proponents claim that a prevailing *erroneous* belief in offense dominance was the master cause of World War I, as it best explains how a small crisis triggered a spiral of hostility and war that all sides wished to avoid. If the participants of World War I had recognized the objective defensive advantages of new technological advances in rifles, machine guns, and artillery, according to one offense-defense scholar, "they would have rushed for their own trenches rather than for the enemy's territory."[4] Scholars have examined the effects of the offense-defense balance not only on interstate war but also on ethnic and civil conflict, arms control, arms racing, alliance behavior, crisis behavior, military doctrine, the consequences of revolutions, the number and size of states and empires, grand strategy, and the structure of the international system.[5] Moreover, the theory offers potentially useful policy prescriptions for reducing competition and preventing conflict in the world today.

A Flawed Theory

This book contends that offense-defense theory is flawed. The theory offers misleading analysis and, regarding its prescriptions, false comforts. The basic problem with the theory is that it mistakenly views technology as a powerful and largely autonomous cause of war and peace. In contrast, this book argues that although technological changes clearly matter, the impact of these changes is always filtered through the strategies that state decision makers employ in pursuit of their political goals. In simplest terms, politics—more than technology—is the master. Moreover, in regard to the content of political goals, offense-defense theory rests on the unsound assumption that most states favor the status quo most of the time and would be happy to live in peace if the nature of technology permitted. In contrast, this book argues that most states face strong incentives to gain power at the expense of others even when technological conditions appear to offer a high level of security.

My critique targets the two principal causal claims of offense-defense theory. First, the theory holds that in any given circumstances, an objective offense-defense balance helps determine *military outcomes*. In other words, proponents claim, the balance plays an important role in shaping the success or failure of military operations on the battlefield. This book shows that changes in the offense-defense balance of technology have little influence on combat outcomes. There are few cases in modern history—perhaps in the late nineteenth and early twentieth centuries (with the great advances in small arms and artillery) and the latter half of the twentieth century (with the development of nuclear weapons)—where technological innovation had a discernible impact on the relative ease of attack and defense. However, in these cases the objective (or, in the case of nuclear weapons, logical) causal effects of technology on combat outcomes are complex and do not conform well to the claims of offense-defense theory. Most often, technological developments have had minimal effects on the relative success or failure of attack and defense. This basic indeterminacy results from the interaction of multiple technologies at different levels of warfare and the far greater influence of other variables that underpin military capability, such as the relative number and quality of the opposing forces.

The second causal link in offense-defense theory is that political leaders' subjective perceptions of shifts in the offense-defense balance help determine *political outcomes*. According to the theory, perceptions of the balance play an important role in state security behavior and, especially, national decisions to initiate war. "Actual offense dominance has been rather rare," according to a main proponent of the theory, ". . . but perceived offense dominance is pervasive, and it plays a major role in causing most wars."[6] This book shows that although leaders usually "get it right" when seeking to understand the military effects of new weapons and innovations, these perceptions do not influence military planning and political decision making as predicted by offense-defense theory. In fact, in none of the cases examined in this book did the perceived offense or defense dominance of technology play an important part in decisions to initiate conflict. Beliefs in the offensive or defensive advantages of military technology appear to do little to either dampen or promote conflict. In sum, offense-defense theory does not fit the facts.

There is an additional causal claim underpinning offense-defense theory, which holds that states can both more easily signal their peaceful intentions and more quickly recognize the aggressive intentions of other states when offensive and defensive weapons are distinguishable. If weapons could be distinguished, according to the theory, states interested in preserving the status quo could credibly convey their benign intentions to other friendly states by deploying defensive forces and limiting offensive forces. Deploying defensive forces under conditions of offense-dominance would send an especially reassuring signal.[7] Correspondingly, states that deploy offensive forces, especially under conditions of defense dominance, would trigger alarms and lead status quo states to take steps to deter aggression. In short, wars arising from miscalculation, misperception, or uncertainty could be significantly reduced if information about political intentions were conveyed by the nature of military deployments. However, this book's critique of the first two causal links in offense-defense theory undermines the claim about states' ability to signal benign intentions (or recognize impending belligerence) through the offense-defense nature of force postures. The root problem is that neither scholars nor policymakers have discovered any meaningful way to distinguish between offensive and defensive weapons.[8] Thus, it is notoriously difficult to derive accurate political intentions from the offensiveness or defensiveness of force postures. Consider one example: Offense-defense proponents typically argue that Britain and France in the 1930s perceived that defense was dominant (based on the lessons of World War I).[9] Yet the majority of leaders in these countries retained their view that Adolf Hitler could be conciliated despite the fact that he was rebuilding the German army in part through tanks and, it was incorrectly believed, heavy bombers. The problem is that these were just the kinds of weapons that offense-defense theory predicts should have triggered alarms—why would Germany build offense in a defense-dominant world unless it had the objective of conquest?[10] If weapons cannot be distinguished in offense-defense terms, their deployment will not inherently convey information about a state's aggressive or benign intentions. Indeed, this book finds no evidence of attempted benign signaling behavior by states.

An Alternative Explanation: Technological Opportunism

The historical cases examined in this book were selected because they offer the greatest leverage for testing the empirical validity of offense-defense theory. In presenting the evidence, however, I also assess the plausibility of an alternative realist explanation for the role of technological change in international politics. This explanation, which I label "technological opportunism," rejects the idea that particular configurations of technology play an important role in ameliorating insecurity in the international system. Technological opportunism, which is derived from the logic of offensive realism, holds that states operating in an anarchical world fraught with uncertainty are compelled to improve their power position in the international system.[11] This basic imperative to maximize relative power is compounded by the fact that leaders face enormous uncertainty when it comes to evaluating the offense-defense balance and predicting technological change. Even if leaders could measure an offense-defense balance, they would have trouble determining how much defense dominance was enough to preserve national security at any given time. Thus, states will rarely view technological developments as a means to maintain the status quo or preserve their power position. Even purportedly defensive technological advances will tend to be seized upon by states as potential opportunities to pursue offensive political objectives. In short, from the perspective of technological opportunism, technology is neither a panacea for nor a fount of war, but something that states employ in pursuit of given policies.

The contrast offered by technological opportunism comes clearly into focus when one considers the two causal claims of offense-defense theory discussed above. First, in terms of explaining *military outcomes,* technological opportunism contends that the relative distribution of power—which includes the balance of military forces (i.e., troops and weapons) and differences in skill in employing those forces (i.e., strategy, doctrine, tactics)—is a better predictor of combat outcomes than the offense-defense balance. To the extent that technology does play a role in determining winners and losers in battle, the impact arises when one state has a technological edge over an adversary based on the array of weapons at its disposal, not when the nature of technology favors either offense or defense. In short, relative power, not the relative ease of attack and defense, better explains success or failure on the battlefield.

Second, in terms of explaining *political outcomes,* technological opportunism views the endemic uncertainty about the current and future impact of technology as more important than perceptions of the offense-defense balance. However, the point should not be taken to mean that technology plays a decisive role in national decisions about whether to initiate war. After all, technological opportunism holds that states are inherently primed for offense and rightly fearful of other states' intentions, and thus will typically find in technological change either cold comfort or new opportunities for gaining power relative to others. Regardless, the nature of technology is a subsidiary factor in shaping decisions to launch war. To paraphrase a like-minded critic of offense-defense theory, where political will and strategic effort are strong enough, war will find a way.[12]

Technological opportunism's basic view that the inherent unpredictability of technological change intensifies the international imperative for maximizing relative power is nicely illustrated by the U.S. decision to build the hydrogen bomb in 1950. Although the United States had built and used atomic (fission) bombs in 1945, American leaders were caught by surprise by the Soviet Union's atomic test in 1949, a technological development that came much sooner than most had expected. A majority of top American officials and advisers argued that exponentially more powerful hydrogen (fusion) bombs were unnecessary because an expanded atomic arsenal would be sufficient to deter a Soviet attack *even if* the Soviets acquired the hydrogen bomb. Moreover, these officials argued, U.S. pursuit of the hydrogen bomb would only encourage the Soviets to match American efforts, leading to an unprecedented arms race and intensification of anxiety around the world. However, the view that ultimately prevailed was that the United States had to build the hydrogen bomb because it needed to regain nuclear superiority, was uncertain about whether the Soviets were already building a hydrogen bomb of their own, and feared that a Soviet hydrogen bomb capability would embolden the Soviet Union to take new risks and seek additional gains.[13] As U.S. Secretary of State Dean Acheson and Secretary of Defense Louis Johnson advised President Harry S. Truman, "sole possession by the Soviet Union of this weapon would cause severe damage not only to our military posture but to our foreign policy position."[14] American leaders were simply unwilling to pass up the opportunity to extend strategic superiority to a new realm of technology, regardless of the potentially destabilizing effects. In short, technological opportunism offers a plausible explanation for why states appear not to be satisfied with the "stability" of defensive technology and are inclined to bolster their capabilities without fear of exacerbating security competition.

This book's analysis of the relationship between technology and the politics of war and peace has important consequences for both political science scholarship and foreign policy. The inextricable link between the realms of theory and policy in this project should be manifest, not least because the contending theories under consideration emerged out of real world policy debates among those seeking to make sense of the cold war nuclear arms competition. In academia, offense-defense theory underpins the well-known and widely employed concept of the security dilemma, lies at the heart of the main debate within contemporary realist theory, and continues to influence a diverse range of scholarship. In the policy world, basic offense-defense ideas guide modern arms control efforts, yield potentially useful prescriptions for reducing interstate competition and preventing civil war, and may shed light on the wisdom of current national security and global nuclear nonproliferation and counterterrorism policies.

Impact of Offense-Defense Theory on Scholarship

Offense-defense theory emerged out of continuing efforts to understand the causes of war and international competition that occupied scholars during the cold war. To be sure, offense-defense explanations have deeper roots. Allusions to the relative strength of attack and defense, or even the idea that offensive advantages foster conflict whereas defensive advantages promote peace, can be traced back to the much older writings of Sun Tzu, Jean-Jacques Rousseau, Carl von Clausewitz, and Antoine-Henri Jomini. (For example, Clausewitz wrote that "the greater strength of the defensive [might] tame the elemental fury of war"; after all, he continued, "if the attack were the stronger form . . . no one would want to do anything but attack.")[15] Moreover, in the last century, basic offense-defense insights are found in the analytical work of B. H. Liddell Hart and J. F. C. Fuller in the 1930s; Marion William Boggs and Quincy Wright in the 1940s; and Malcolm Hoag and Thomas Schelling in the 1960s.[16]

The foundational works on offense-defense theory, however, were produced in the 1970s and 1980s at a time when cold war strategic nuclear issues loomed large in academic and policy debates. International relations scholars such as George Quester, Robert Jervis, Shai Feldman, Jack Snyder, Stephen Van Evera, and Charles Glaser believed that nuclear weapons gave defenders a large military advantage, ensured strategic stability in the form of mutual deterrence, and should have allowed U.S. and Soviet leaders to feel tremendously secure and act more peacefully. Offense-defense proponents thus saw the superpower nuclear arms race as driven by a misunderstanding of the nature of the nuclear revolution and feared that the unnecessary competition could spiral into an otherwise avoidable nuclear war.

In this context, the originators of offense-defense theory sought to identify more general conditions under which military technology might cause war or peace.[17] The best-known work in this vein is Robert Jervis's 1978 article "Cooperation under the Security Dilemma." The security dilemma holds that many of the steps pursued by states to bolster their own security simultaneously make other states less secure.[18] States are compelled to interpret other states' military preparations as hostile, even when such preparations are meant solely for self-defense, because international anarchy (the absence of a central authority to govern or protect states) and pervasive uncertainty about present or future intentions and the shifting balance of power makes inaction inordinately risky. This triggers an action-reaction spiral of reciprocal arms buildups, diplomatic tension, and hostility that can lead to war. Even if war is avoided, the security dilemma and resulting spiral model can leave all states worse off than they were when they started, if simply by creating a more dangerous security environment or wasting valuable resources on an unnecessary arms race.[19] Leaders' misperceptions and misunderstandings of one another's actions and motivations exacerbate the security dilemma, but it is important to note that the power of the concept lies in its rational foundation. The security dilemma shows how states that are relatively satisfied with the status quo, mainly concerned with national security, and primarily taking defensive measures can wind up in an unintended, undesired, and unnecessary conflict.[20]

The security dilemma results in a vicious circle of intense power competition that sometimes ends in war. The roots of this dilemma lie in the anarchic structure of the international system, but anarchy is a fixed condition, not a variable. How then can one explain varying periods of war and peace, instability and stability, conflict and cooperation? More generally, is intense security competition among states an inevitable consequence of the structure of the international system?

Competing answers to this question account for the main divide within contemporary realism, which remains the single most important approach for understanding international politics.[21] The contending answers offered by the two main schools—defensive realism and offensive realism—largely turn on competing views of the utility of offense-defense theory.[22] In a nutshell, defensive realism holds that the strong propensity of states to compete is not an inevitable consequence of the structure of the international system because sometimes the nature of technology allows states to feel safe, demonstrate their commitment to the status quo, and signal their benign intentions. In other words, offense-defense variables provide a systematic method of predicting when the security dilemma is severe and when it is not. Offensive realism, on the other hand, sees the security dilemma as an immutable feature of international politics so long as states operate in anarchy. For offensive realism, any variation in state security behavior is explained not by offense-defense variables but by the opportunities and constraints created by a constantly shifting balance of power.

The realist tradition rests on a set of common assumptions and premises: the international system is anarchic, meaning that there is no central authority or governing power to protect states from one another; states are the most important actors in world politics; states seek to survive and generally act rationally in pursuit of this and other goals; uncertainty and the shadow of military force are central to international politics; and the competition for power or security is an enduring feature of international relations. Offensive and defensive realists part ways on the issue of how much power states desire and are willing to pursue in the name of security. Defensive realists accept the idea that anarchy, uncertainty, and the shadow of violence compel states to care deeply about their security and the aggressive intentions of other states. However, defensive realists argue that the pursuit of excessive power in a benign security environment is likely to leave a state worse off in the end because this behavior will tend to provoke a hostile balancing coalition or unnecessary arms racing. Absent a threatening environment, states can safely pursue policies aimed at preserving the existing balance of power. To be sure, none of this precludes the possibility that aggressive states will pursue hostile expansionist policies regardless of the security environment. However, when the security dilemma is less severe, moderate and sometimes even cooperative behavior among most states can be the best route to security in international politics.[23]

According to defensive realism, the severity of the security dilemma depends on whether offense or defense has the advantage on the battlefield (i.e., the offense-defense balance) and whether offensive and defensive capabilities can be distinguished.[24] When the balance favors offense, the probability of competition and war is greater for several reasons. First, striking first is an attractive option for all states. If technology is such that states believe an initial attack will lead to a quick and decisive victory, then striking first will be tempting—both because of fear of attack under such conditions and the desire to make gains through conquest. Second, political crises are more likely to escalate into full-fledged military conflicts because states will be quicker to conclude they are threatened, more inclined to preemptively attack, and more prone to accidental attacks.[25] Third, arms races become more intense because even small advantages in armament levels can have decisive consequences in a short war. Finally, when offense has the advantage, international politics become more rigid, polarized, and destabilizing because states must recruit allies well in advance of a conflict.[26] In sum, all types of war—e.g., expansionist, preemptive, preventive, accidental—are more likely when the balance favors offense. The opposite holds when defense has the advantage.

Defensive realism also holds that war is also more likely when weapons and military postures that protect the state are indistinguishable from those that provide the capability for attack. The ability to differentiate between offense and defense allows status quo states to behave in ways that are clearly different—and more reassuring—from those of aggressors. Most important, as long as the offense-defense balance does not greatly favor attackers, offense-defense distinguishability allows states to both protect their own territory and signal their peaceful intentions by deploying defensive forces, thereby minimizing unnecessary conflict and competition. Even purely opportunistic aggression is less likely when offense and defense can be distinguished because states are better able to recognize any offensive arms buildup and take countermeasures to deter attack. Finally, offense-defense differentiation creates the option for states to limit or ban weapons that favor attack while allowing those that favor defense, thereby bolstering deterrence and security. As Jervis concludes, "The advantage of the defense can only ameliorate the security dilemma. A differentiation between offensive and defensive stances comes close to abolishing it."[27]

Offensive realism contends that as long as states function in an anarchical world, in which all states possess some offensive military capability and can never be certain of one another's current and future intentions, they are compelled to maximize their relative power position.[28] The reason is simple: the more power a state has relative to others, the greater its chances of survival. States are thus always looking for opportunities to expand their power. In this sense, all states are essentially revisionist because they desire to revise the existing distribution of power in their favor. Mutual security is simply not attainable and thus not pursued. For these reasons, offensive realism rejects the idea that the security dilemma could ever be eliminated.

This book shows that offense-defense variables face serious logical, operational, and empirical problems. If states are unable to measure the offense-defense balance or distinguish between offensive and defensive forces—or if offense-defense variables have little actual impact on military or political outcomes—then the security dilemma cannot be ameliorated and states cannot convincingly signal their benign intentions.

Offensive realism does recognize that sometimes the costs and risks of expansionist behavior outweigh its benefits, which explains why international politics is not in a constant state of war. But this has little to do with offense-defense variables. Instead, the circumstances in which opportunistic aggression pays depend on the distribution of power in the international system. Specifically, offensive realism predicts that security competition, expansionist behavior, and war are more likely when there are more than a few great powers in the world and when power is unevenly distributed among the great powers. Larger numbers of great powers increase the likelihood that these states will seek to pass the buck to one another when it comes to opposing aggression, whereas uneven distributions of power tempt stronger great powers to pursue hegemony over weaker states. Regardless of variations in the distribution of power, much less offense-defense variables, offensive realism contends that states are fundamentally insecure and tend not to pass up opportunities to dominate one another whenever they arise.

In sum, offense-defense theory allows defensive realists to argue that intense security competition is not an inevitable consequence of the structure of the international system. In an offensive realist world, any variation in security competition flows from the relative distribution of power among states. For defensive realists, the degree to which one state threatens another is a function of the distribution of power filtered through the offense-defense balance. If power is distributed roughly equally and the balance of technology does not heavily favor offense, states may not only feel secure with defensive strategies and forces but also signal their peaceful intentions to other status quo states through such defensive military postures.

As a final testament to the wide-ranging and enduring scholarly influence of offense-defense theory, it is worth noting the theory's role in the literature on the microfoundations of international conflict, including approaches as diverse as game theoretic crisis bargaining and social constructivism.[29] Consider James Fearon's oft-cited article "Rationalist Explanations for War," which addresses the puzzle of why rationally led states are sometimes unable to locate a negotiated settlement that all sides should prefer to fighting a risky and costly war. Fearon argues that one of the few valid causal logics by which war can occur under these conditions arises from the unresolvable commitment problem inherent in preemptive war scenarios caused by offensive technological advantage.[30] Alexander Wendt, in his seminal *Social Theory of International Politics,* focuses on the question of how states can create a culture of trust, avoid excessive egoism, and achieve global peace under the condition of international anarchy. Wendt argues that when defensive technology has a significant and known advantage, states are constrained from going to war and may be willing to trust one another enough to start the process of collective identity formation. In essence, the offense-defense balance may provide a functionally equivalent substitute for an externally constraining central authority. Wendt cautions that the process of creating an international culture of trust will generally need to move beyond external constraints (such as defensive advantage) and rely on internal constraints (such as self-restraint in behavior toward other states). However, his examples of how states might credibly allay one another's security fears by imposing visible sacrifices on themselves—for example, by unilaterally giving up certain technologies—show that offense-defense ideas play a multidimensional role in Wendt's constructivist analysis.[31]

To be sure, these examples and much of the additional scholarship employing offense-defense variables simply assume that offensive advantages are more dangerous than defensive advantages without devoting much attention to the theoretical and empirical foundations of offense-defense theory. The point here is not to criticize these works but merely to note that although only a few scholars have offered a systematic analysis of offense-defense theory, many experts of international security have used basic offense-defense ideas in their research.

Policy Significance of Offense-Defense Theory

The idea that the nature of military technology affects the prospects for war and peace looms large in international relations scholarship, but it also has clear policy significance. Proponents of offense-defense theory have occasionally exaggerated claims of policy impact—for example, that "Mikhail Gorbachev's Soviet regime . . . absorbed offense-defense theory, and between 1985 and 1991 shifted toward a defensive military doctrine partly because of it."[32] However, offense-defense ideas provide the foundation for modern disarmament and arms control policies and can potentially shed light on vital contemporary security issues, such as the emerging revolution in military affairs, missile defense, nuclear proliferation, and terrorism.

Technology has been harnessed in the service of war since the Stone Age, but it has also been looked to by statesmen for more peaceful purposes, especially in the last century, when the destructiveness of modern technology rose exponentially. After World War I, scholars and diplomats began to advance the idea that offensive weapons might foster war and defensive weapons might encourage peace. The diplomatic negotiations at the 1932 World Disarmament Conference sought to classify, limit, and eventually abolish "aggressive" armaments.[33] In his address to the conference, United States President Franklin Delano Roosevelt declared, "[I]f all nations will agree wholly to eliminate from possession and use the weapons which make possible a successful attack, defenses automatically become impregnable, and the frontiers and independence of every nation will become secure."[34] The conference failed to achieve its basic goal, but attempts to identify and limit offensive capabilities as a means to prevent war persisted. During World War II, a prominent study noted that "since some weapons are more useful than others in the tactical offensive, it is clear that . . . a regulation of weapons may have an influence on the capacity of the political offensive to advance itself by resort to arms."[35]

The idea of limiting weapons that favor offense while allowing those that favor defense underpins modern arms control theory, which emerged in the late 1950s and early 1960s and guided arms control efforts throughout the cold war. Indeed, Thomas Schelling, the main developer of modern arms control theory, derived his views through an analysis of how strategic instability created by the search for offensive nuclear advantage could lead to inadvertent war. The basic insight was that even hostile states should see the necessity, virtue, and common interest in limiting the nuclear arms race and preserving nuclear deterrence.[36] At the conventional level, the Western European peace movement that emerged in the 1980s promoted "nonprovocative" military postures, or "nonoffensive defense," as a means of stabilizing the military balance between the North Atlantic Treaty Organization (NATO) and the Warsaw Pact.[37] The concept of nonprovocative defense shaped arms control negotiations at the 1989 Conventional Forces in Europe Treaty, survived the end of the cold war, and, according to some, remains relevant today in deterring, conducting, and settling wars.[38]

Many factors tend to undermine arms control efforts. One of the most significant limitations is the difficulty of identifying and restricting only "provocative" weapons and forces, even leaving aside the obvious political penchant for labeling "offensive" anything in an adversary's arsenal. Consider several important anomalies in arms control policy. The Land Mines Convention of 1997 outlaws possession and use of antipersonnel land mines, even though land mines are viewed as essentially defensive weapons in most military formulations. Similarly, the proliferation of small arms and light weapons is commonly identified as a cause of civil and ethnic conflict and thus the target of preventative arms control measures, even though these weapons are typically seen as an obstacle to tactical assault.[39] In both of these cases, however, advocates of arms control probably conclude that the expected reduction in civilian casualties in civil wars outweighs any benefits of stability.

Competing views of the military and political implications of the spread of ballistic missile technology, deployment of national and theater missile defenses, and the proliferation of nuclear weapons often turn on offense-defense assumptions. For example, ballistic missiles have traditionally been thought of as a relatively offensive weapon and thus destabilizing, which partly explains the current arms control efforts to promote the Missile Technology Control Regime. But missiles can also be viewed as disproportionately advantageous to the defense and thus peace-enhancing. As one military analyst argues, "Ballistic missiles are made to destroy bases. They can disarm an opponent before he can move to an offensive position. It is nearly impossible to engage in military operations where incoming warheads are bursting."[40] From this perspective, Chinese or North Korean concerns about U.S. offensive intentions in sharing theater missile defense with Taiwan and South Korea make sense. The best example, of course, is the U.S. deployment of national missile defense (NMD). Opponents of NMD argue that even if the U.S. seeks to defend solely against the emerging missile capabilities of "rogue states," the system will undermine Russian and Chinese nuclear retaliatory capabilities and possibly fuel a classic spiral of exaggerated hostility.[41] Finally, offense-defense ideas underpin opposing views on whether nuclear proliferation makes the world more or less peaceful. If, as many argue, the nuclear balance of terror helped keep the peace between the superpowers during the cold war, then the spread of nuclear weapons to new states should deter war and stabilize international politics.[42]

In short, the logic of offense-defense theory pervades policy debates in the areas of nuclear and conventional strategy, doctrine, deterrence, and arms control. However, one should not confuse the prominence of offense-defense ideas in policy analyses with the actual influence of these ideas on policymakers. As is shown in chapter 5, in the case of U.S. nuclear strategy and force posture during the cold war, offense-defense views were ignored or dismissed by those making the policy decisions, and contemporary arguments about the destabilizing consequences of national missile defense appear to be falling on deaf ears within the U.S. government. Likewise, no policymaker today would publicly advocate the controlled proliferation of nuclear weapons to states not already in the nuclear club, and few would argue even in private that the benefits of nuclear proliferation outweigh the costs and risks.

Offense-defense theory is on firmer ground in claiming to offer *potentially* useful arms control policies. The most promising avenues would be to help prevent war by either deliberately shifting the balance of technology toward defense or seeking to correct misperceptions of the balance.[43] When the offense-defense balance favors offense, political leaders should cooperatively adopt defensive military force postures and seek arms control agreements to limit offensive forces. This type of policy cooperation increases international security because it not only allows states to signal their nonaggressive intentions but also helps to shift the offense-defense balance toward defense. When the balance favors defense, leaders should also seek to correct any misperceptions of the nature of the balance, thereby eliminating false security fears. In short, offense-defense theory suggests that policymakers may be able to ameliorate dangerous beliefs and behavior through arms control.

Debates about the future of warfare are currently dominated by a focus on the impact of technology. Many claim that advances in information technology and long-range precision striking power are creating a revolution in military affairs (RMA) in which the relationship between offense and defense is being completely transformed. Defense analysts typically argue that the RMA has shifted the offense-defense balance toward offense and created strong preemptive strike incentives for states that have mastered the new technologies. These predictions of offensive advantage and preemption came well before the United States launched its astonishingly quick and decisive campaigns in Afghanistan (2001) and Iraq (2003), and even before many of the newest precision strike technologies were on display in Kosovo (1999).[44] Moreover, the continued development of long-range precision-strike capabilities could threaten the survival of existing nuclear deterrent arsenals around the world and thus destabilize nuclear deterrence.

The last important policy topic upon which offense-defense ideas bear directly is terrorism.[45] There are at least two important dimensions. First, as Richard Betts argues, there may be an operational advantage of attack over defense in the interactions of terrorists and their opponents. If the offense-defense balance does indeed favor offensive terrorist operations, the United States and its allies should emphasize counteroffensive operations—including preemptive and preventive attacks—rather than homeland and other passive defenses to defeat them.[46] Second, the specter of nuclear terrorism undermines the relative optimism of offense-defense theorists' understanding of the nuclear revolution. Many believe that nuclear deterrence is the functional equivalent of defense dominance—it is hard to imagine a state in possession of a secure nuclear arsenal being attacked or invaded by another state—but deterrence is unlikely to prevent suicidal terrorists from using nuclear weapons if they have the means to do so. The implication is that the "defensive" effects of nuclear weapons come into play only in interstate relations and there is less reason to believe that the next sixty years will be as free of nuclear warfare as the previous era.

In sum, offense-defense ideas have broad application in theory and practice. A thorough evaluation of offense-defense theory and its main alternative is relevant to a large and growing body of scholarly research in international relations and will help shed light on past and emerging foreign policy debates.

Testing the Theory

In 1984 Jack Levy reviewed the first phase of offense-defense scholarship and concluded that "the concept of the offensive/defensive balance of military technology needs more theoretical attention and operational definition before it can be applied to systematic empirical analysis."[47] Levy's survey found logically flawed or incomplete hypotheses; multiple, confusing, and contradictory definitions of the offense-defense balance; and inconsistent historical classifications of periods of offense or defense dominance. In the two decades since, proponents of offense-defense theory have sought to address the conceptual and procedural shortcomings identified by Levy.

Is offense-defense theory now developed enough to warrant empirical testing? On the one hand, much progress has been made. Stephen Van Evera's book *Causes of War: Power and the Roots of Conflict* presents a comprehensive treatment of the core hypotheses concerning how offensive and defensive advantages affect the likelihood of war. Van Evera offers the overarching thesis that offense dominance is a taproot cause of war. He delineates an extensive list of war-causing mechanisms created by offense dominance (most of which center on the nature of diplomacy, arms racing, and incentives to initiate attacks), qualifies the theory by discussing the rare conditions under which offensive capabilities promote peace, outlines the factors that shape the offense-defense balance, infers predictions from the theory, and evaluates these predictions based on a measure of the offense-defense balance over the last two hundred years.[48] Other offense-defense theory proponents, including Sean Lynn-Jones, Charles Glaser, and Chaim Kaufmann, have sought to refine and develop the theoretical foundations of the theory with an eye to formulating testable hypotheses.[49] On the other hand, the theory still faces important logical problems, the core concept of an offense-defense balance has remained difficult to define and identify in practice, and rigorous tests of offense-defense hypotheses have been surprisingly rare. (Chapter 1 addresses these problems in depth.) Indeed, almost twenty years after Levy's survey, a review essay by Richard Betts found that most of the attempts by proponents to refine offense-defense theory had succeeded only in muddying the conceptual waters and undermining the theory's claim to analytical and practical utility.[50]

Fortunately, there is ample justification for putting offense-defense theory to the test. For one thing, offense-defense proponents themselves are united in the belief that scholars can and should push the envelope of empirical analysis. Lynn-Jones encourages students of the theory to "formulate testable hypotheses and to test them empirically," while Glaser argues that "the full range of offense-defense hypotheses warrants further empirical testing."[51]

A second reason why an empirical evaluation of offense-defense theory is justified is that it is possible to derive an operationally testable version of the theory based on careful deductive reasoning and attention to the shared understandings and assumptions of offense-defense proponents. One of the dangers in testing an underdeveloped theory is that a critic might force the theory into making predictions beyond its scope and lead to an easily refutable caricature or "straw man" version of the theory. This book seeks to avoid that charge by taking offense-defense theory on its own terms and deriving only the most clear and precise predictions and causal mechanisms that flow from the core logic of the theory.

A third reason to believe that lingering conceptual obstacles can be overcome is that scholars have begun to undertake increasingly sophisticated empirical evaluations of offense-defense theory. These empirical assessments have employed a range of social science research methods and differing levels of analysis to get at the basic validity of offense-defense hypotheses. The results of these studies have been widely divergent and sometimes contradictory. For example, one recent statistical analysis finds strong support for hypotheses about the effects of offense-defense balance on historical trends in attack and conquest, while another reveals no statistically significant effects of the balance on either the likelihood of war or the onset of militarized disputes that fall short of full-blown war.[52] Similarly, Stephen Van Evera finds strong evidence of the effects of offensive advantage on military and political outcomes in three historical case studies, while Stephen Biddle uses case studies, statistical analysis, and computer experimentation to show that shifts in the balance of technology generated no meaningful change in attacker success rates, thus throwing into doubt the sweeping political consequences claimed by offense-defense theory.[53] In short, as the offense-defense research program enters a new phase of testing, this book offers a comprehensive theoretical and empirical critique.

Finally, offense-defense theory deserves empirical evaluation because it appears to have the attributes of a good theory. If the aim of social science should be to develop accurate and relevant explanations of human behavior and social institutions, then offense-defense theory plausibly fits the bill.[54] First, it addresses a vitally important topic and seeks to develop knowledge that matters beyond academia. Second, the theory offers a simple and compelling explanation. Third, the theory claims great explanatory power, not just in terms of causal significance but also wide range and real-world relevance. Arguably, these attributes render offense-defense theory particularly deserving of greater empirical analysis and evaluation, even if additional theoretical development makes sense.

Research Questions

The prime hypothesis of offense-defense theory is that offensive advantages make war more likely. This prime hypothesis consists of two explanatory hypotheses, one relating to military outcomes and the other to political outcomes. The first explanatory hypothesis is that offensive shifts in the balance make quick and decisive military victories for the attacker more likely. The second is that perceptions of offensive shifts in the balance make leaders more willing to initiate conflict.

Two straightforward research questions thus drive this study. First, is there an offense-defense balance that can be used to predict or explain military outcomes? Second, do perceptions of shifts in the balance affect political decisions to initiate conflict? It is easy to see that these questions get to the heart of offense-defense theory. However, it is much harder to transform these questions into concrete testable predictions because the offense-defense balance concept is exceedingly complex. What observed events should be interpreted as representing a shift in the balance? Chapter 1 analyzes the concept of an offense-defense balance as a means to infer good empirically testable predictions. The chapter presents basic definitions and assumptions underlying the balance, differentiates between "core" and "broad" versions of the balance used by offense-defense proponents, and identifies and evaluates the most frequently employed criteria for classifying how technology gives a relative advantage to offense or defense. Chapter 1 infers and logically defends two testable predictions for the explanatory hypothesis about *military outcomes*. First, mobility-improving technological innovations favor attackers and result in quicker and more decisive warfare. Second, firepower-improving technologies favor defenders and result in longer, more indecisive warfare. The chapter also derives two testable predictions for the explanatory hypothesis about *political outcomes*. First, political and military leaders are more inclined to initiate war when they believe technological change has given offense a large advantage. Second, leaders are deterred from initiating war when they believe technological change has given defense a large advantage. These military and political predictions are the most useful, clearly articulated, and frequently employed predictions in offense-defense scholarship.

Case Study Method

I employ the historical case study method to investigate how these predictions fare against the record of military and political outcomes in periods of great technological change. Below I explain why and how the specific cases were chosen to provide maximum theoretical leverage. More
generally, the case study method is the most suitable methodology for the
task at hand because it allows for the detailed exploration of offense-
defense theory's hypothesized causal mechanisms.[55] Statistical analyses
are more appropriate for estimating the general causal effects of the
offense-defense balance across a large number of cases, whereas the use of
formal models provides a rigorous way to assess the deductive logic of
offense-defense hypotheses prior to empirical testing. However, even
when formal analysis reveals logically consistent causal mechanisms and
statistical analysis validates assertions about causal effects, a complete
causal explanation requires case study analysis to empirically substantiate
whether the proposed causal mechanisms functioned as expected in the
process leading to observed outcomes.[56] In other words, even if a causal
explanation makes sense and the purported cause is associated with the
observed effect, we still need to assess whether the cause actually shaped
the outcome and how the causal process worked.

The case study method is particularly useful in testing offense-defense
theory for several reasons. First, most proponents of the theory acknowledge that the offense-defense balance has strong observable effects on
military and political outcomes only when it takes on an extreme value—
that is, when technological change dramatically shifts the balance toward
offense or defense. These large shifts are relatively rare, however, so scholars face the practical challenge of evaluating offense-defense theory with
a small number ("small-n") of cases. Although the familiar "small n/large
N" distinction is not the most salient difference between case study and
statistical approaches, it is true that quantitative studies require a sufficiently large sample of cases or data set to achieve significant findings,
whereas case study methods can yield decisive results with few cases or
even a single case.[57] Second, case studies make it easier to assess the causes
and effects of variables that are particularly difficult to define and measure, such as the offense-defense balance. Indeed, statistical studies of
offense-defense theory explicitly or implicitly rely on previous case study
research to code historical periods of offense and defense dominance. As
with many variables in the social sciences that are difficult to measure (for
example, power or democracy), measuring the offense-defense balance
and its influence on military and political outcomes requires detailed con-

sideration of contextual factors, which is one of the great comparative advantages of the case study method.[58] A third reason why the case study method is apt for testing offense-defense theory is that there is a comparatively greater need to examine the theory's causal mechanisms (not just its causal logic and effects) and, perhaps, narrow the scope conditions under which these mechanisms are expected to operate.[59] As a structural theory, offense-defense purports to explain overall trends in international politics, but it also yields clear predictions for the behavior of particular states in the wake of major technological changes. A lack of correlation between the value of the offense-defense balance and the expected behavior in these cases does not necessarily point to an inherent theoretical flaw—countervailing effects of other important variables may have negated the effects of the balance—but careful case studies can reveal whether the causal mechanisms identified by offense-defense theory were in fact operating as predicted.

The case study chapters employ congruence and process-tracing procedures to test offense-defense theory. The congruence procedure is a within-case method of causal interpretation in which the predicted values of the independent and dependent variables of a test hypothesis are compared with the observed values of those variables.[60] (For example, were quicker and more decisive combat outcomes consistent with the large increase in mobility wrought by railroads? Were leaders deterred from initiating war at times when they believed technological change had given defense a large advantage?) Ascertaining this congruence (consistency) or incongruence (inconsistency) will not by itself confirm or disconfirm offense-defense theory, but it might indicate a possible causal relationship, especially if the observed outcomes vary in a similar direction and magnitude as predicted, and it creates a framework for process tracing.

Process tracing is a within-case method of seeking to identify the intervening causal process and intervening variables between an explanatory variable and an outcome variable.[61] If a case reveals congruence between predicted and observed values, then one needs to assess whether this correlation is causal or spurious (that is, whether both variables are correlated only because of the presence of a third, antecedent variable). For example, if we observe the emergence of railroads and an increase in quick and decisive warfare, we need to examine whether both were caused by a separate variable, such as the rise of a great continental power among weaker neighbors. Likewise, if the congruence procedure reveals inconsistency between explanatory and outcome variables, process tracing allows one to determine whether there was indeed a causal relationship that was simply masked by the operation of other variables. For example, if railroads and quick and decisive battles were not correlated, we might find that the attacker employed railroads in a suboptimal way or that the defender possessed much greater military power than the attacker, which should not be taken in itself as a failure of the theory. Similarly, beliefs in offensive advantage and aggressive state behavior may not be correlated as predicted because of the influence of other variables (domestic politics, the balance of power, issues of individual character, etc.), but nonetheless there may be a causal process linking such beliefs with a greater willingness to initiate war. Process tracing requires attention to historical context and detail in order to examine whether the causal mechanisms hypothesized or implied by a theory in a case are in fact operating in the sequential steps in the causal process in that case.[62] In short, each chapter traces whether the central causal mechanisms of offense-defense theory—the impact of the offense-defense balance on military outcomes and the role of perceptions of the offense-defense balance on political outcomes—were in fact operative in the case.

Avoiding Selection Bias

Practitioners of the case study method have frequently been warned of the dangers of selection bias, generally understood as the problem that occurs when a researcher deliberately selects cases in a way that results in causal inferences that suffer from systematic error.[63] In a theory-testing project, the concern would be whether the success or failure of a theory's predictions in a small number of cases might be driven by the initial choice of misleading or unrepresentative cases. The selection bias warning frequently centers on the practice of selecting cases where the dependent variable takes only one kind of extreme value—for example, if a theory of war and peace were tested only with cases where war occurred.[64] However, the more relevant concern here is whether foreknowledge of historical outcomes, and the subtle (or not so subtle) desire to confirm or disconfirm particular hypotheses, biases the selection of cases in the first place.

This book avoids the potential danger of selection bias because it provides a clear rationale for case selection that best serves the ultimate research objectives and because its case studies were chosen in a way that puts offense-defense theory on the strongest possible ground. As discussed above, chapter 1 derives the most logical and commonly agreed-upon criteria for identifying shifts in the offense-defense balance: improvements in mobility favor offense, and improvements in firepower favor defense. The next step is to ask, what have been the watershed technological innovations in military mobility and firepower in modern history? The two biggest advances in military mobility are the emergence of railroads in the nineteenth century (which is explored in chapter 2) and the introduction of tanks in the first half of the twentieth century (chapter 4). The two most important advances in firepower are the artillery and small arms revolution of the late nineteenth and early twentieth centuries (chapter 3) and the nuclear revolution of the latter half of the twentieth century (chapter 5). Because the overall research objective is to test whether the offense-defense balance plays the powerful role in determining military and political outcomes claimed by offense-defense theory, it makes sense to select cases in which the balance (the independent variable) takes extreme values. If the balance has real effects, we should witness those effects in the above cases where the balance has purportedly shifted dramatically toward offense or defense. In short, the method of selecting "most likely" or "easy" cases offers a powerful means of theory testing: if offense-defense theory explains international politics anywhere, it should do so under these conditions.[65]

Case Study Findings

How well does offense-defense theory fare in these cases? Does technological opportunism appear to be more consistent with the outcomes and offer a more promising avenue for further research? The case study chapters compare the relevant offense-defense predictions with military and political outcomes. Did major mobility innovations—railroads (chapter 2) and tanks (chapter 4)—shift the offense-defense balance toward offense and make quick and decisive victories for the attacker more likely? Were leaders more inclined to initiate war when they believed the balance had shifted in favor of offense? Did major firepower innovations—small arms and artillery (chapter 3) and nuclear weapons (chapter 5)—shift the balance toward defense and result in longer and more indecisive wars of attrition? Were leaders more inclined to feel secure and act benignly when they perceived defense to be dominant?[66]

Chapter 2 explores the impact of railroad mobility on the relative ease of attack and defense on the battlefield and the role that perceptions of the offense-defense capability of railroads had on political decisions to initiate conflict. In terms of military outcomes, railroads made armies much more mobile than ever before and were employed in several quick and decisive wars at the time. The evidence shows, however, that although attackers profited from greater strategic mobility conferred by railroads, this benefit was more than offset by the defenders' gains from improved operational mobility. Regardless, the rapid and stunning victories that occurred in Europe at the time—namely, Prussia's defeat of Austria in 1866 and France in 1870–71—resulted primarily from great asymmetries in military skill, not because of offensive advantages in technology. In terms of political outcomes, the evidence directly contradicts offense-defense predictions. Leaders were more than willing to initiate conflict even when railroads were perceived to favor the defender. If anything, war occurred less frequently when railroads were thought to favor the attacker.

Chapter 3 investigates the impact of the firepower revolution in small arms and artillery on the offense-defense balance and the relationship between perceptions of technology and war initiation. The evidence shows that these advances in firepower made attacks on prepared defenders much more difficult. Firepower shifted the balance in favor of defenders, as offense-defense theory predicts. This finding is qualified, however. The defensive advantages lay primarily at the tactical level of warfare, and doctrinal choices (not just technological conditions) significantly shaped the problem of offense at the time. More problematic for the theory is the evidence regarding political outcomes. Contrary to conventional wisdom, but according to new evidence, German leaders before the outbreak of World War I were well aware that technological advances in small arms and artillery had made offense extremely costly and difficult and that a coming war was likely to be long and protracted. Yet these accurate perceptions of the causes and effects of the offense-defense balance did not foster stability or deter aggression.

Chapter 4 analyzes the impact of tank technology on military operations and political behavior. In terms of combat outcomes, evidence from the key period of World War II, when adversaries were equally skilled at armored warfare, shows that tank mobility had an indeterminate effect on the offense-defense balance. Most important, in terms of political outcomes, Adolf Hitler chose to attack his neighbors absent any belief that tanks favored offense over defense.

Chapter 5 examines the effects of the nuclear revolution on war and politics. The logic used by offense-defense proponents to code nuclear weapons as enormously defense-dominant may be internally consistent, but it is also based on political—not solely military—calculations. Nevertheless, the fact that war between nuclear-armed states—or at least between great power nuclear-armed states—has not occurred supports offense-defense theory. Once again, however, the bulk of the evidence undermines offense-defense predictions about political behavior. Leaders in the United States and the Soviet Union during the cold war understood that nuclear weapons had rendered warfare prohibitively costly, yet the two countries engaged in an intense and costly arms race, built highly offensive strategic forces, and intervened often in the third world. Enormous defense dominance did not appear to make these states feel secure and act benignly.

The book's conclusion summarizes my principal findings. Recall each of the two general questions that dominate this study. First, is there an offense-defense balance of technology that can be used to predict military outcomes? Proponents of offense-defense theory have often offered ambiguous and problematic definitions of the offense-defense balance, struggled to provide objective and consistent criteria for distinguishing between offense- and defense-enhancing technologies, and failed to provide proper guidance for measuring the offense-defense balance at any given time. The logic of the offense-defense balance is arguably sound, if highly complex, but the importance of the balance in conditioning military outcomes is not supported by the empirical evidence. The effects of new technologies are rarely so powerfully advantageous to either offense or defense that they demonstrably shape battlefield outcomes—if simply because any effects are washed out by factors such as the balance of military forces, comparative fighting skill, strategic planning, and military doctrine.

Second, do perceptions of the offense-defense balance affect political decisions to initiate conflict? Here offense-defense theory ultimately falls far short as a useful theory of international politics. In cases where leaders had generally correct perceptions of the offense-defense balance, their strategic behavior contradicted offense-defense predictions. Concerns about offensive or defensive advantages were overshadowed by more significant and tangible strategic and political factors. In none of the cases examined in this study was the perceived offense dominance of technology an important factor in decisions to initiate war.

Technological opportunism offers a more promising avenue of research. At best, the case studies explored here can only be considered a "plausibility probe," but the empirical evidence is more consistent with the predictions of technological opportunism than offense-defense theory. Regardless of whether or not leaders were able to characterize shifts in the offense-defense balance, these leaders typically seized upon new technologies as potential opportunities to gain military and political advantage over their rivals. For example, German leaders before World War I were well aware that new firepower technology and weapons had made frontal assaults by infantry more difficult. Yet these leaders were not deterred from launching war. To the contrary, German strategists sought ways to capitalize on the advantage of defensive firepower in the conduct of offensive military operations. The case of U.S. nuclear weapons policy during the cold war may offer even greater support for technological opportunism. American leaders did not think and act in ways predicted by nuclear defense dominance, but technological opportunism will need to be tested directly against other explanations that see American behavior as driven by massive misperceptions of the requirements of deterrence, organizational imperatives, or domestic politics.

This book argues that international military and political outcomes are largely independent of an offense-defense balance of technology. The findings are bound to disappoint anyone attracted to offense-defense theory for its apparent elegance, intuitive appeal, explanatory power, or normative relevance. Nevertheless, the theory does not, contrary to the words of one offense-defense proponent, "fit the way things work, or the way we think."[67]

1

The Offense-Defense Balance

The label "offense-defense theory" applies to a body of scholarship that explores the causes and effects of variations in the offense-defense balance. What is the offense-defense balance? What causes it to change? As with any well-established research program, in this literature one can find important conceptual differences and divergent claims but also a common set of assumptions and predictions. For example, virtually all proponents of offense-defense theory view technology as the main variable shaping the relative ease of attack and defense, just as all are united in seeing a greater threat to international peace and stability arising from offense dominance.[1] This chapter discusses the commonly accepted definitions, understandings, and assumptions underpinning the offense-defense balance; it differentiates between the "core" and "broad" versions of the balance; and it argues that an approach that seeks to apply the core version of the balance offers a more useful and promising research design. The chapter then deduces the clearest possible criteria for attempting to measure the offense-defense balance. Based on the criteria most commonly employed by proponents of offense-defense theory, two hypotheses are inferred for how to classify the impact of technological change on the relative efficacy of offense or defense. These classification criteria are then used as a guide to select historical cases, which are used to test offense-defense predictions about military and political outcomes in subsequent chapters.

The Offense-Defense Balance

The offense-defense balance is the central theoretical construct of offense-defense theory, but scholars working on the subject have conceptualized, operationalized, and employed the balance differently.[2] This diversity notwithstanding, several central features of the balance are common to almost all approaches.

Shared Assumptions

The offense-defense balance is defined as the relative ease of attack and defense given prevailing conditions. Relative ease denotes some measure of the relative costs and benefits of attack and defense. Offense and defense refer to the use of military force, not the political motivations, intentions, or goals that may motivate such military action. *Offense* is identified as the use of force against another state in order to seize territory or destroy assets. *Defense* entails the use of force to block those attacking forces.

Offense and defense can be used to describe military action at three levels of warfare: strategy, operations, and tactics. The strategic level of analysis pertains to overall war plans and ultimate outcomes; the operational level concerns the conduct of specific campaigns in a war, and the tactical level concerns actions and engagements taken within a campaign or battle.[3] In order to measure the offense-defense balance and its effects on military and political outcomes, the strategic level would appear to be the most suitable level of analysis. Because the theory ultimately aims to explain the politics of war and peace (i.e., decisions to initiate war), what matters most is leaders' expectations of whether attackers can win wars quickly and decisively. After all, launching a war would not be appealing to leaders if the nature of the balance promised victory in the opening campaigns but defeat in the war. In practice, however, the operational level of warfare offers a suitable unit of analysis, especially for measuring the objective offense-defense balance. For one thing, the success or failure of specific campaigns will very often determine whether the attacker or defender prevails in a given war. And, given the great uncertainty military and political leaders face in predicting the course of any war very far into the future, the prospect that initial campaigns can be won rapidly serves as a good functional equivalent for the prospect of strategic success. The same point is less applicable for translating tactical outcomes to operational outcomes. On the one hand, the relative feasibility of offensive and defensive tactics should play a role in operational outcomes—how else are campaigns won but through individual combat engagements? On the other hand, any operation typically depends on both offensive and defensive tactics to achieve success. For example, successful offensive military operations to conquer territory almost always entail tactical defensive actions to hold newly seized territory. Correspondingly, a defending force at the operational level usually depends on the tactical offensive to regain lost territory.[4] In sum, the logic of offense-defense theory implies that the strategic offense-defense balance is the key unit of analysis, but in practice the offense-defense balance at the level of operations is what should drive military and political outcomes.[5]

Proponents of the theory have struggled to offer a more precise definition of the balance than just the relative ease of attack and defense.[6] Perhaps the most frequently deployed definition of the balance is cast in terms of a cost or investment ratio required for offensive success: that is, the amount of resources (e.g., dollars) an attacker must invest in offensive capability to offset the amount of resources a defender invests in defensive forces.[7] This definition is problematic because it renders the offense-defense balance unobservable in practice and in principle. To measure any given offense-defense balance defined in terms of a cost ratio, one would need to do so in three steps. First, one would need to differentiate what counts as money spent on offense from money spent on defense, which depends on the ability to classify weapons systems as either—mostly? partly? entirely?—offensive or defensive. Second, one would need to determine what attackers and defenders actually spent on these systems in a given historical case. Third, one would need to calculate counterfactually what each side needed to spend to offset the other side's investment. In other words, one would need to estimate "how much more a war's loser *would have had* to have spent in order barely to prevail—or how much less the winner *could have* spent and still barely have won."[8] If the first two steps are intractable (at best), the third step requires one to calculate an unobservable quantity. Fortunately, for present purposes even the operationally problematic definition is useful because it demonstrates that the offense-defense balance is a continuous variable, not a dichotomous one. What matters most for empirically evaluating the impact of the balance on military and political outcomes is not whether the balance in any given period favors offense or defense in absolute terms—in fact, it is almost always easier (less costly) to defend than to attack—but how and to what degree the balance has shifted in either direction.

There is a near consensus in the literature that the offense-defense balance must be distinguished from two other relevant variables in international politics: power and skill. First, the balance must be defined independently of the distribution of power among states. Battlefield outcomes clearly depend on a host of factors other than the offense-defense balance, including the balance of military forces and resources among the adversaries. Therefore, the success of any given offensive or defensive strategy is not necessarily indicative of the offense-defense balance. For example, if a given state easily conquers another state, it might be misleading to conclude that offense is relatively easier than defense because that outcome could simply have resulted from the attacking state's numerically or qualitatively superior forces rather than from an offensive advantage. Similarly, if only one state in a war has acquired a new technology or weapon, and this unilateral advantage is largely responsible for that state's success in battle (as either attacker or defender), we can only safely conclude that the outcome resulted from a change in the balance of power, not necessarily in the offense-defense balance. The effect of a new technology on the balance is *ideally* assessed based on battlefield evidence from a conflict between two roughly equal-sized military forces deploying similar weapons and technologies. Of course, the effects of the offense-defense balance can hypothetically overcome even large disparities in power in determining war outcomes, but the offense-defense balance and the balance of power must be treated as analytically and empirically distinct.

Second, the offense-defense balance in any given circumstances should be defined independently of large disparities in the level of military skill between opposing forces. The offense-defense balance is broadly meant to capture something about the objective effects of military technology on war and politics. To isolate and interpret these baseline effects, one must imagine or locate cases in which states make reasonably rational decisions about force posture, doctrine, and strategy.[9] The standard of optimality employed by proponents does not require that attackers and defenders make the absolute best strategic choices (assuming such a determination is even possible). Instead, optimality in this context means determining the offense-defense balance when states make reasonably intelligent decisions about how to employ existing technologies and forces given prevailing knowledge at the time.

Core and Broad Versions of the Offense-Defense Balance

There are two schools of thought on the determinants of the offense-defense balance. The "core" version looks almost exclusively at how changes in military technology shift the relative advantage of attack and defense.[10] Two types of change in the existing pool of technology matter. First, technological change may produce an entirely new type of weapon or combination of weapon systems that makes it less costly to conduct offensive or defensive operations. Second, nonmilitary technological progress may result in more efficient methods for producing, improving, or employing given weapons or systems (for example, through scientific advances in metallurgy, electronics, or computing, or through the commercial development of mass production or information management systems).[11] Precisely how technology favors offense or defense will be discussed in the next major section. In addition to technology, however, offense-defense explanations sometimes incorporate other factors that purportedly shape offensive and defensive advantage. Factors included in the "broad" version of the offense-defense balance include geography, the cumulativity of resources (i.e., the ease of exploiting resources from conquered territories), nationalism, regime popularity, alliance behavior, force size, and military doctrine, posture, and deployment. Incorporating these factors into the balance makes an already complex concept even harder to operationalize, but proponents of the broad version contend that the core version is otherwise incomplete. The relative ease of attack or defense is determined by a set of basic causal factors, they argue, so omitting any of these factors might result in flawed explanations or predictions.

Geography   Geography affects the offense-defense balance in several ways. First, geographical barriers to movement such as oceans, lakes, rivers, mountains, and forests favor defenders over attackers. Terrain or man-made features of territory (such as urban sprawl) that provide cover reduce the speed of attack and thus give a relative advantage to defense. Other geography-related factors that aid defense include large distances between attackers and defenders, which increase the cost of projecting military force; wide buffer regions between attackers and defenders, such as third states or demilitarized zones that may be off-limits to attacking forces; rural settings, which facilitate guerrilla resistance against invaders; and insulated resources or industries, which make economic strangulation tougher for the attacker.[12]

Cumulativity of Resources   The cumulativity of resources is the degree to which resources can be exploited from conquered territories. According to offense-defense theorists, the balance shifts toward offense when resource exploitation is easier because it reduces the cost of offensive operations. The easier it is for an attacker to exploit a defender's natural resources or productive effort of the population after invasion, the easier and cheaper conquest will be.[13]

Nationalism   According to proponents of offense-defense theory, nationalism tends to favor defense relative to offense for at least two reasons. First, people are more likely to fight harder when they believe they are defending their rightful homeland from foreign invaders. In these cases, offense is harder than it would be if a defender's nationalist claims did not coincide with its actual borders. Second, a nationalistic citizenry has a "wartime extraction advantage," meaning that a defender can rely more on the productive energies of its (besieged) people than can an attacker. Knowledge of this wartime extraction advantage allows the defender to invest less in defense before the outbreak of a war. However, offense-defense proponents also note two exceptions when nationalism may favor attackers: "Where there are significant irredenta . . . nationalists may fight just as hard on offense as on defense"; and "where multinational empires rule disgruntled subject people, nationalism may actually favor offense by raising the probability that invaders will not be resisted but welcomed as liberators."[14]

Regime Popularity   Popular regimes may give either offense or defense the advantage in warfare. One scholar writes, "Popularity of regime probably aided offense before roughly 1800 but aided defense since then." Before about 1800, popular regimes could raise larger and more loyal armies that were able to operate far from the attacker's homeland, thus aiding offensive power. After 1800, popular governments could better organize and arm their citizens for guerrilla resistance against invaders, thus aiding the defense. Unpopular regimes at any time are more vulnerable to subversion, which is characterized as a form of offense.[15]

Alliance Behavior   Even proponents of the broad version of the offense-defense balance disagree on whether international alliance behavior should be considered a cause of the offense-defense balance. Some scholars argue that collective security systems, defensive alliances, and balancing behavior by neutral states favor defenders because aggressors will face greater opposition than in the absence of these diplomatic arrangements. If states bandwagon (i.e., join the stronger coalition against the weaker one), offense is favored, because aggressors gain allies as they expand.[16] Others exclude alliance behavior on the grounds that including it is inconsistent with the central goal of offense-defense theory—to explain state behavior based on more or less exogenous constraints—and creates a problem of circularity or even tautology (if the balance is both an effect and a cause of alliance behavior).[17]

Force Size   Advocates of the broad version of the theory believe that force size affects the offense-defense balance but disagree on how. One approach argues that force size may aid either offense or defense. According to this view, large, mass armies generally aid the offense, but "once armies grow so big that they can cover an entire frontier . . . their size aids the defense because offensive outflanking maneuvers against them become impossible."[18] Another approach holds that defense becomes more advantageous as force size increases. This argument is supported with examples from both conventional and nuclear war. First, larger force size may allow defenders to achieve adequate force-to-space ratios (i.e., the size of forces in relation to the length of the front) to successfully defeat an offensive at any force-to-force ratio. Second, because a successful nuclear offensive depends on the ability to eliminate a defender's retaliatory capability, increases in defender force levels make success increasingly less likely.[19]

Military Doctrine, Posture, and Deployments   Proponents of the broad theory are also divided on whether military doctrine, posture, and deployments shape the offense-defense balance. Stephen Van Evera argues that military doctrines, especially those concerning the use of a new technological innovation, can change the balance. Offense became easier, for example, when the blitzkrieg doctrine made motorized armor a more effective instrument in the attack. Van Evera also contends that states shape the offense-defense balance by their military posture and force deployments. In particular, forces deployed in vulnerable positions render offensive operations against those forces easier. Van Evera writes, for example, "Stalin eased attack for both himself and Hitler during 1939–41 by moving most of the Red Army out of strong defensive positions on Soviet territory and forward into newly seized territories in Poland, Bessarabia, Finland, and the Baltic states. This left Soviet forces better positioned to attack Germany, and far easier for Germany to attack, as the early success of Hitler's 1941 invasion revealed."[20] Charles Glaser and Chaim Kaufmann, on the other hand, argue for excluding military doctrine, posture, and force deployments because the theory assumes states act optimally. As they argue, "When states act optimally, doctrine and deployments merely reflect the balance; they are outputs of the optimization process, given the constraints imposed by the offense-defense balance and the distribution of resources. Suboptimal choices will influence a state's deployed capabilities but not the offense-defense balance."[21]

Why the Core Version Is Better

Although proponents of the broad version of the offense-defense balance believe their approach strengthens offense-defense theory, there are several important theoretical and practical advantages to focusing solely on the core balance of technology. First, the fact is that technology is the one determinant of the balance common to all versions of the theory, it is the most significant factor shaping the balance, and it is often the only factor analyzed in any detail by scholars. None of the factors identified by the broad approach have such wide applicability and importance. Thus offense-defense theory's contribution to the conceptual toolbox of international relations turns largely on the role that the technological balance plays in shaping state behavior. Moreover, all things being equal, the relative parsimony of the core version makes the theory easier to operationalize and measure and renders it more intuitively appealing.

Second, because the broad offense-defense balance incorporates factors unique to particular states (such as geography, nationalism and regime popularity), it is not a systemic variable, and the resulting theory is no longer structural. This matters because offense-defense theory claims to share the appeal of other structural theories in international relations because it focuses on the war-causing effects of a variable that is essentially exogenous to states. Technology, in principle, provides similar constraints and opportunities for all states in a given international system.[22] Comparable patterns of state behavior should thus arise under similar technological balances. At best, the broadly defined balance might shed light on a specific military conflict, but its effects are not generalizable across time.

Finally, and most fundamentally, the broad version of the balance ultimately renders offense-defense theory atheoretical and nonfalsifiable. The problem stems not just from a lack of parsimony, which complicates theory testing. Rather, the problem with adopting the broad version of the balance is that doing so transforms the theory from a general causal statement with an explanation (that is, one that describes and explains the cause and effect of a class of phenomena) into a grab bag of variables employed in a post hoc descriptive enterprise for each specific and unique case. The factors identified in the broad version may shape to some degree the relative ease of attack and defense. More broadly, the outbreak of violence in any single case results from an inevitably complex set of opportunities and constraints, motives and goals, and decisions and actions. In short, almost everything matters to some extent when addressing the causes of social phenomena. However, explanations of military and political outcomes that rest on a laundry list of variables are not only intuitively unappealing but essentially nongeneralizable and nontheoretical. The core version of the balance, on the other hand, rests on an interesting theoretical claim: there is something about military technology itself that affects the likelihood of war and peace across time and place.

An explanatory approach based on the broad balance is also nonfalsifiable. The broad balance consists of a list of complex, sometimes ambiguous, and often crosscutting variables. A falsifiable multivariate approach requires the stipulation of clear and objective criteria for weighing the relative causal importance of each explanatory variable. In other words, one needs guidance not only for how to accurately assess the relative impact of each of the broad factors on the offense-defense balance but also on how to aggregate these effects into a single value of the balance.[23] Proponents of offense-defense theory provide no such criteria, and thus a judgment about which of the variables on a list matter most in any given case is likely to seem arbitrary.[24] The danger here is that the absence of theoretical guidance for weighing multiple variables can lead analysts to pick and choose among the list of explanations based on their foreknowledge of how well those explanations fit the evidence in any given case. In short, theoretical predictions stemming from the broad conception of the balance can be neither empirically supported nor refuted with much confidence because of the inherent elasticity of the variables shaping the balance.

In sum, this book's focus on the central variable of technology is wise because the core version of the offense-defense balance provides the only promising avenue for offense-defense scholarship. Technology is the most common and significant factor shaping the balance. Technology is, in principle, a systemic variable that can be used in a structural theory of international politics. The core balance of military technology can arguably be operationalized and measured, whereas the broad balance cannot. The core version of offense-defense theory identifies a single variable that is used to explain complex phenomena, whereas the broad version relies on all possible causes to explain the same phenomena. In short, the core theory is satisfying, whereas the broad approach is not.

The Effect of Technology on the Offense-Defense Balance

What are the criteria used to identify how technology gives a relative advantage to offense or defense at any given time? Without such coding criteria, we have no theoretical guidance for judging which factors contribute disproportionately to offense or defense on the battlefield and thus cannot assess the causal effects of the offense-defense balance. This section sets forth and evaluates the logic of two criteria: (1) mobility-enhancing technologies favor offense; (2) firepower-enhancing technologies favor defense. Proponents have generally struggled to provide objective and consistent criteria for distinguishing between offensive and defensive changes in technology. The few explicit discussions of differentiation criteria have often been supported by ambiguous arguments or contradictory examples. Despite a high degree of confusion and the fact that not all offense-defense proponents make these claims explicit, the mobility and firepower criteria are the most useful, clearly articulated, and frequently employed hypotheses about the causes of the offense-defense balance.

Critics of offense-defense theory often argue that the offense-defense balance cannot be measured because weapons cannot be categorized as offensive or defensive.[25] Proponents respond that the balance, properly understood, does not depend on the ability to label weapons as inherently or entirely offensive or defensive, but only on whether any given weapon at a particular time makes offense or defense easier. As one proponent writes, "Individual weapons systems almost invariably combine technologies that can be labeled offensive or defensive."[26] The disagreement may be mostly semantic or strictly technical because, in practice, proponents frequently do classify weapons as almost entirely offensive or defensive. For example, the same proponent writes of the "virtually unambiguous offensive character of bridging equipment, which is mainly useful for enabling attacking forces to cross rivers inside enemy territory."[27] Regardless, the point proponents would stress is that what matters for assessing the balance is not individual weapons per se but the general technological characteristics underlying a pool of weapons systems at any given time. How can one best label these technologies?

Hypothesis 1: Mobility-Enhancing Technologies Favor Offense

Almost all proponents of offense-defense theory believe that new or improved technologies that enhance the mobility of military forces contribute relatively more to offense than defense. This section examines previous historical and contemporary offense-defense discussions of mobility as a characteristic of offense and evaluates the logic behind the mobility-favors-offense hypothesis.

Early Views of Mobility and Offense   The 1932 World Disarmament Conference unsuccessfully sought to distinguish and abolish offensive ("aggressive") weapons based on the technological characteristic of mobility. Critics of offense-defense theory have seized on this failure to argue that the goal of categorizing offensive and defensive weapons is a hopeless one. For proponents, the source of the problem at the conference was each country's desire to prevent cuts in its own forces, not the inability to identify offensive and defensive weapons.[28] The debate cannot be settled here, but a brief discussion of the conference negotiations shows both the inherent obstacles to classifying weapons technology and the kernel of a mobility-favors-offense hypothesis.

Consider first the diplomatic negotiations over three categories of land forces in which mobility was an issue: artillery, tanks and armored vehicles, and fortifications. Proposals to abolish certain types of artillery based on their mobility frequently arose but "were so hedged about with reservations as to prevent any practical action."[29] One group of states believed that all tanks were offensive, another group of states viewed only heavy tanks as offensive, and a third group of states insisted that only tanks capable of assaulting modern fortifications of medium strength were offensive. Complicating the tank question was the fact that the delegates were unable to distinguish between armored vehicles and tanks and, in any event, could not determine whether armored vehicles were offensive or defensive in the first place.[30] Finally, the conference could not even agree that fortifications, so far as they served as an obstacle to mobility, favored defenders. Germany, for example, argued that French forts were offensive because their location so close to the Franco-German border meant that the forts could be used as a base to launch an attack on Germany.[31]

The effort at offense-defense classification in the negotiations on naval and air forces was equally unfruitful. In the area of naval weapons, the conference studied the possible offensive nature of capital ships, aircraft carriers, and submarines but failed to reach any conclusions. As one observer writes, "The United States and Great Britain, with great battle fleets, asserted the defensive character of capital ships and aircraft carriers, but regarded the submarine, the principal naval weapon capable of threatening the supremacy of surface sea power, as offensive. Powers with lesser navies took the view that the main components of the battle fleet were aggressive, but submarines were defensive." The conference was also unable to classify any airplane as offensive. Although the commission studying the issue leaned toward labeling bombers as offensive, it could only conclude rather cryptically that airplanes "regarded as most efficacious against national defense are those which are capable of the most effective direct action by the dropping or launching of means of warfare of any kind."[32]

One analyst of the conference proceedings, Marion Boggs, does conclude that it might be possible to identify offensive "potentialities" underlying actual weapons and forces. Boggs writes, "The offense possesses mobility and striking power, and protection to a lesser degree . . . Movement or mobility hence emerges as a significant criterion of distinction . . . 'and mobility is essentially an offensive quality.' Armament which greatly facilitates the forward movement of the attacker might be said tentatively to possess relatively greater offensive power than weapons which contribute primarily to the stability of the defender."[33]

Two influential commentators on military affairs in the interwar period, J. F. C. Fuller and B. H. Liddell Hart, also developed important ideas on the offense-defense impact of weapons technology. The two clashed on the utility of classifying weapons for the purpose of qualitative disarmament. Fuller argued that it was meaningless to label any weapon "aggressive" because all weapons could be used aggressively, and thus it would be absurd to try to abolish weapons based on their inherent characteristics. Liddell Hart, on the other hand, believed that certain weapons were indispensable to aggressors, and thus war was less likely if these weapons were banned. Although Fuller and Liddell Hart disagreed on this issue, they did agree that certain weapons and technologies could contribute relatively more to offense. "Some weapons undoubtedly possess a higher offensive power than others," Fuller acknowledged. Likewise, Liddell Hart referred to "the kind of weapons which inherently favoured the offensive."[34] At times, both saw mobility as the key characteristic of an offensive weapon. Fuller believed that mobility had always been the primary factor in the offensive: "As long as armies are small enough to maneuver freely, and are commanded by generals with an equal mobility of mind . . . offensive power will be high."[35] Fuller thus became the leading proponent of the idea that mechanization and armored warfare were the key to offensive success and had the potential to revolutionize war. He argued that if tanks (and airplanes) were properly employed to maximize mobility, victory for the attacker could be quick and decisive.[36] Liddell Hart also viewed mobility on the battlefield as the most important feature of successful offensives. He "championed the view that the tank provided the means to restore mobility to the battlefield and make it possible, once again, to win quick and decisive victories," and referred to his new "concept of reviving the power of the offensive and the art of war, by lightning strokes with highly mobile mechanised forces." By the late 1930s, however, Liddell Hart had reversed his opinion and now viewed mobility as even more important for the defense, as it allowed for the redeployment of forces to threatened points along a front.[37] He argued that "despite the apparent advantage that mechanization has brought to the offensive, its reinforcement of the defensive may prove greater still."[38]

Contemporary Views of Mobility and Offense   Contemporary scholars of offense-defense theory have devoted far more effort to theorizing about the consequences of changes in the offense-defense balance than to explicating how weapons technologies actually determine the balance. Given the current prominence of offense-defense theory, this paucity of coding criteria is surprising. Consider how several proponents have sought to describe the technological foundations of offense.

George Quester, in his broad survey of warfare in Western history that serves as one of the founding works of offense-defense theory, aims to identify the factors "that make it advantageous to strike out offensively at the enemy, rather than to sit in prepared defensive positions waiting for him to strike."[39] Quester is clear on these factors in at least one case: "Mobility thus generally supports the offensive. First, one can invade with impunity if one can bring along all the "comforts of home," all of one's most deadly vehicles of destruction. Second, the ability to move may allow an attacking force to exploit various weak spots or blind spots of the force that is standing in place . . . Third, the ability to move allows an attacking force to group itself, to assemble temporary numerical superiorities as it pleases, when it decides to begin battles."[40] Quester's own historical examples, however, sometimes contradict this mobility-favors-offense argument. For example, Quester notes that the appearance of the mounted warrior in the fourth century strengthened the offense because horses greatly improved mobility, but then suggests that the emergence of the mounted knight in the tenth century strengthened the defense. His argument is that the knight was dependent on the castle from which he drew his financial support, thus limiting the range within which the knight was willing to fight. Note, however, that this argument is not based on technological characteristics.[41]

In his seminal piece on offense-defense theory, Robert Jervis reveals pessimism about the ability to concretely define or identify the offense-defense variables that shape the severity of the security dilemma. He provides little criteria for determining when offense has the advantage or for identifying offensive weapons and technologies. Jervis writes that ground warfare "is a contest between fortifications and supporting light weapons on the one hand, and mobility and heavier weapons that clear the way for the attack on the other . . . [However], there is no simple way to determine which is dominant."[42] Even a complex method for determining the offense-defense balance is elusive if, as Jervis acknowledges, the balance can vary over the course of a single war, campaign, or battle. (Unlike few proponents before or since, Jervis attempted to assess the conventional offense-defense balance at the time of his writing in 1978. However, he concludes that he is "unable to render any firm judgment.")[43] The problem of distinguishing offense from defense is even worse: "No simple and unambiguous definition is possible and in many cases no judgment can be reached." Jervis ultimately concludes that "whether a weapon is offensive or defensive often depends on the particular situation—for instance, the geographical setting and the way in which the weapon is used."[44] Judging from his above statement about fortifications versus mobility, as well as other statements (cited below) about the characteristics of defensive systems, one might justifiably assume that Jervis believes mobility generally favors offense.[45]

Sean Lynn-Jones defines offensive technologies as those that "make it less expensive for states to seek security by adopting offensive military postures and strategies," but he does not indicate what forces contribute to offensive military postures and strategies in the first place.[46] Lynn-Jones explains,

> The pool of available technologies at any given time determines the cost of building weapons and deploying military capabilities that can be used in support of an offensive or defensive strategy. Silo-based ICBMs, for example, combine a "defensive" technology—the hardened silo—with an "offensive" one—multiple, highly accurate, highly destructive nuclear warheads. The net result, according to offense-defense theory, is that states can deploy invulnerable retaliatory forces at relatively low cost, making conquest virtually impossible.[47]

From this one might conclude that firepower favors offense. Elsewhere, Lynn-Jones refers to cannons, siege machinery, and tanks as offensive, suggesting that the underlying technology common to all three—firepower—is typically an offensive characteristic.[48] On the other hand, tanks are usually seen as powerful offensive weapons because they combine firepower with protective armor in a mobile vehicle. In sum, it is difficult to derive from Lynn-Jones's work explicit criteria for identifying the technological characteristics of offense.

In one of the definitive works on offense-defense theory, Stephen Van Evera writes that "military technology can favor the aggressor or the defender," but he provides no criteria for making such a determination.[49] The lack of explicit criteria for how military technology favors attackers leads Van Evera to make what seem to be ad hoc or even contradictory claims. His discussion of the impact of mobility on the offense-defense balance is one illustration. Van Evera seems to conclude, unlike most proponents, that mobility favors the defense: "In modern times, technology that gave defenders . . . greater mobility . . . strengthened the defense." Similarly, "technologies that favored chariot or cavalry warfare . . .

strengthened the defense." He also argues, however, that "revolutionary France's mass armies strengthened the offense because they had greater mobility."[50] Each of these statements could be plausible descriptions of historical developments; however, without general criteria for identifying which technologies favor offense one cannot operationalize and measure the effects of offense-dominance on military or political outcomes or verify the empirical descriptions advanced above.

Among proponents of offense-defense theory, only Charles Glaser and Chaim Kaufmann offer an explicit discussion of how given technological characteristics might shift the balance towards offense.[51] "The offense-defense impact of a specific weapons or technology innovation cannot be assessed simply by considering its performance properties in isolation; rather, we must assess its impact on states' abilities to perform offensive and defensive missions."[52] How can one assess this impact? Glaser and Kaufmann write, "The most critical question in this process is how the innovation differentially affects advancing forces and nonadvancing forces. Innovations that are usable only or primarily by nonadvancing forces will tend to favor defense, while innovations that are equally usable by forces that are advancing into enemy-controlled territory will favor the offense."[53] According to the authors, improvements in mobility favor offense because they multiply the attacker's advantage of the initiative. That is, greater mobility increases the attacker's ability to quickly outflank or overwhelm a surprised defender and reduces the time an attacker must take to assault defensive positions.[54]

Evaluating the Mobility-Favors-Offense Hypothesis   The above discussion indicates a consensus view that mobility-enhancing technologies favor offense, even if the underlying causal logic is not often explicated. A brief discussion of that logic lends credibility to that claim. In military terms, mobility is the ability of troops and equipment to move from one place to another. There are essentially three types of mobility: strategic, operational, and tactical. Strategic mobility is the ability to transport military forces from the homeland to a theater of operations or between theaters. Offense-defense theorists suggest that greater strategic mobility allows the attacker to expeditiously transport and supply its forces far from its own borders, thus negating the defender's geographic advantage. Operational mobility is the ability to move forces within a theater. According to proponents, greater operational mobility allows the attacker to concentrate forces quickly to achieve a numerical advantage on a small portion of the front, rapidly exploit weak points in a defender's line, or outflank a defender's position altogether. Tactical mobility is the ability to move forces on the battlefield, in the face of enemy fire. Offense-defense proponents argue that greater tactical mobility reduces the number of casualties suffered by an attacker because these losses are partly a function of the amount of time that forces are exposed to enemy fire in an assault.[55]

There are several counterarguments to the mobility-favors-offense explanation. First, in terms of strategic mobility, it is not obvious why the ability to transport and supply forces far from the homeland gives an attacker much of an advantage over a defender who already has this capability. Once an attack is under way, in fact, the defender depends more than the attacker on the ability to quickly move forces to that theater. Moreover, the impact of strategic mobility appears indeterminate when the defender relies more heavily than the attacker on reinforcement from overseas territories and allies. Second, in terms of operational mobility, one could argue that the attacker depends more on the element of surprise than on mobility to achieve a successful breakthrough of a defender's frontline, whereas the defender places a premium on mobility to reinforce threatened points in the front. Unless a breakthrough, penetration, or envelopment occurs so rapidly that the defender never has a chance to react and counterattack, the defender would also seem to profit more from mobility once an advance is under way.[56] Third, greater tactical mobility may be more advantageous to the defender than the attacker in several ways. Tactical mobility allows defenders to trade space for time through a series of tactical withdrawals to fortified positions where they can continue to fire on attacking forces. In addition, greater offensive tactical mobility may actually increase attacker casualties, as the greater speed of an assault often comes at the price of reconnaissance, protection, and preparatory artillery fire.[57] Finally, tactical mobility is advantageous for the defender because the defender often must seize the tactical counteroffensive to avoid defeat.

On a very abstract level, the argument that mobility favors offense seems plausible. Imagine a world of two states, for example, each inhabiting a separate island with no means of transportation over water. One could say that the lack of mobility favors defense in this case because offense is impossible. Moreover, the invention of a canoe in this world would seemingly contribute relatively more to offense than defense simply because offense would no longer be impossible. Beyond a simple and abstract case such as this, however, the logic of the mobility-favors-offense explanation becomes murkier. To better illustrate how mobility might favor defenders over attackers, consider a competing analogy from American football. An effective offense must have speedy running backs and receivers. Yet just as important for the offense is initiative—the ability of running backs to pick a hole in the line to run through and the ability of receivers to make a sudden cut while running a pattern downfield. An effective defense must have speedy linebackers, cornerbacks, and safeties that can react quickly as offensive plays develop. Mobility would appear to favor defense over offense because linebackers depend on greater quickness to prevent running backs from breaking through the defensive line, and cornerbacks and safeties rely on greater speed to close the gap between them and receivers before thrown balls arrive. These competing analogies notwithstanding, the mobility-favors-offense criterion is the most widely accepted proposition among offense-defense proponents.[58]

Hypothesis 2: Firepower-Enhancing Technologies Favor Defense

The next most useful criterion for assessing the offense-defense balance is offered by the claim that technological innovations that enhance firepower favor defense. Early thinkers and modern proponents of offense-defense theory are less explicit on the role of firepower than on the role of mobility, but the most cogent arguments show that the firepower-favors-defense hypothesis offers a clear criterion for judging the impact of military technology on the balance.

Views on Firepower and Defense   In his analysis of the 1932 World Disarmament Conference, Boggs writes, "The defense disposes especially of striking power and protection, to a lesser extent of mobility."[59] This characterization of "striking power," or firepower, as the key element of defense is also noted by other early offense-defense scholars, including Quincy Wright in his classic study of war.[60]

Quester, in his discussion of the criteria for distinguishing defensive technologies, writes that "weapons become supremely defensive whenever a foreign decision to violate frontiers is indispensable to providing them with a target." Similarly, he argues, "any weapon that relates to peculiarities of terrain will be supportive of the defense." On these grounds, Quester classifies improvements in fortification technology and the building of walls, garrisons, and castles as essentially defense-enhancing. He later acknowledges, however, that the net impact of fortifications can favor offense in at least two ways. First, fortresses can be exploited to hold one frontier, thus freeing forces to attack on another frontier. Second, fortresses "can also support the mobility of highly maneuverable offensive forces, by reducing the amount of food and ammunition they must carry with them as they race from front to front."[61] Quester's views on firepower are similarly ambivalent. On the one hand, he discusses how the introduction of the longbow, musket, and cannon tremendously strengthened offense over defense. On the other hand, he argues that more modern advances in firepower favored defense. Quester notes that seventeenth-century artillery restored a defensive advantage: "Artillery after all is not useful only for battering down walls. If mounted properly within walls, it can perhaps cannonade the besieging forces before they succeed in getting their artillery into place. Once fortified structures were redesigned to exploit artillery . . . the balance could again begin to shift to the defense." Ultimately, Quester's discussion of the *mis*perceived advantages of offense before World War I indicates his agreement with the firepower-favors-defense hypothesis.[62]

Jervis writes, "Although there are almost no weapons and strategies that are useful only for attacking, there are some that are almost exclusively defensive." He cites "total immobility" and "anything else that can serve only as a barrier against attacking troops" as characteristics that define a purely defensive system.[63] Although the latter example might be interpreted as saying that anything exclusively defensive is defensive, the former example indicates Jervis's belief that mobility generally favors offense, while barriers to mobility favor defense. In terms of firepower, Jervis argues that light weapons, machine guns, and nuclear weapons favor defense, whereas heavy weapons and artillery favor offense.[64]

Van Evera's views on firepower are unclear. On the one hand, he notes that lethal small arms (such as fast-firing rifles) favor the defense. On the other hand, he argues that technologies that favor the mass production of small arms strengthen the offense.[65] Glaser and Kaufmann, once again, provide the most clear-cut analysis for why improvements in firepower favor defense: "In battle, attackers are usually more vulnerable to fire than are defenders because they must advance, often in plain sight of defenders, making them easy to detect and to hit, whereas defenders are often well dug-in and camouflaged."[66] Furthermore, the authors argue, the need to concentrate forces to achieve a breakthrough makes the attacker more vulnerable than defenders to area-effect firepower.[67]

Evaluating the Firepower-Favors-Defense Hypothesis How sound is the firepower-favors-defense hypothesis? Firepower is a measure of the destructive power of the weapons or array of weapons available to adversaries in a conflict. Firepower consists of not only explosive power but also range, accuracy, and rate of fire. There are several good reasons to argue that technological innovations that enhance firepower capability are disproportionately advantageous to the defense. First, firepower allows the defender to threaten the attacker's concentration of forces before an attack. An attacker typically needs a local advantage of combat power to pierce the defender's forward defenses. Numerical superiority requires density, but the greater density of forces provides more targets for defensive fire and thus more attacker casualties. Second, firepower favors defense because it reduces the mobility (i.e., offensive power) of the attacker. In the face of greater defensive fire, an attacker must seek more armored protection, cover, concealment, and dispersal—all of which slow the attacker's advance. Finally, defensive firepower forces the attacker to provide its own covering fire in the advance, which slows the attack because of added weight and time required to reposition covering fire.

There are several reasons to believe, however, that firepower is as crucial in the attack as it is in defense. First, the attacker relies heavily on suppressive or covering fire to neutralize or inhibit defender forces, weapons, or reconnaissance. Suppressive fire by an attacker reduces the amount of fire faced by advancing forces and can pin down defender forces until they can be overrun and destroyed. Second, most successful offensives require preparatory bombardments before the attack. Preparatory barrages can shatter the morale of defenders, destroy defensive positions, and disrupt defender reinforcements and communication. Third, just as the defender uses firepower to disrupt attacker concentrations of forces before an attack, the attacker depends on firepower to disperse defender forces into greater depth away from the front line. Because firepower, especially artillery, does the greatest damage to forces that are grouped together, dispersal is the wisest option. However, when defenders disperse, the force-to-force ratio shifts in the attacker's favor and makes offensive breakthroughs more likely. Finally, and perhaps most important, firepower allows the attacker to selectively thin out his forces elsewhere in order to concentrate forces at the main point or points of attack. In other words, the attacker can substitute firepower for manpower at points on the front where holding the line is the main objective and can use those forces to achieve a numerical superiority at a chosen point where breakthrough is the goal.[68]

In a simple hypothetical case the hypothesis that firepower favors defense seems valid. Consider the previous example of two island states equipped with canoes. The introduction of rifles or cannons into this world might appear to favor the defending island because the attacker, advancing across open water, is more vulnerable to enemy shot. On an actual battlefield, however, things are more complicated. Although an attack in the face of heavy defensive fire is enormously difficult, an offensive undertaken without preparatory and/or suppressive fire is virtually impossible, and a numerically superior concentration of forces would be unlikely, or unwise, without firepower available to strengthen the areas thinned out for the attack. In short, there are good reasons for arguing that firepower is as crucial and useful for defenders as for attackers.

In sum, the mobility and firepower criteria are not unproblematic, but they are plausible hypotheses that have often been used to differentiate offensive advantages from defensive ones. Proponents of offense-defense theory might emphasize that whether these technological characteristics are more useful in the attack or defense is a question not easily considered in logical isolation. Instead, they might argue that the mobility-favors-offense and firepower-favors-defense propositions derive from battlefield evidence and should ultimately be assessed against the historical record.

Have major mobility-improving innovations in the past created the conditions for quicker and more decisive victories for the attacker? Have major firepower-improving technologies made warfare more prolonged and indecisive, thus making such victories less common? Have perceptions of either category of innovations ultimately made war more or less likely? These questions are evaluated in the following chapters.

2

The Railroad Revolution

Railroads were the quintessential technological innovation of the industrial age. The combination of steam engine locomotives rolling on iron rails marked the beginning of a revolution in communication, transportation, and military strategy. The first practical locomotive appeared in the late 1820s. Railroads spread rapidly across Europe and the United States in the 1830s, 1840s, and 1850s, and by mid-century all the major European armies had conducted field exercises in moving and supplying troops by rail. Most of the great powers in the world had built extensive national railroad networks by the 1870s.

Railroads dramatically increased the strategic mobility of armies. The mobilization, deployment, and concentration of ever-larger forces could now be achieved across vast distances at up to ten times the speed of marching troops. Offense-defense theory makes two predictions about the consequences of such significant improvements in mobility. First, greater mobility conferred by the railroads should make quick and decisive victories for the attacker more likely.[1] Second, the tantalizing prospect of quick victory offered by the railroads—or fear that an adversary will seize the same opportunity—should make leaders more inclined to launch war.

This chapter examines the impact of railroads on military and political outcomes between 1850 and 1918. The two research questions are straightforward: Did railroads shift the objective offense-defense balance toward offense? Did perceptions of an offensive advantage of railroads make war a more appealing option? The chapter first discusses the role of railroads in shaping military outcomes in Europe in the 1850s, North America during the Civil War, and Europe during the wars of German unification and World War I. As might be expected, the offense-defense impact of railroads during this period of rapid technological change is complex. Although railroad mobility revolutionized the ability to assemble and concentrate forces for an attack, this advantage was offset by the defender's ability to use railroads to reinforce threatened points and retreat when necessary. In fact, contrary to offense-defense predictions, the bulk of the empirical evidence suggests that railroads favored strategic defenders over attackers.

The chapter then examines how leaders perceived the military significance of railroads at the time and how these perceptions affected their strategic behavior. The available empirical record about beliefs and decision making contradicts the predictions of offense-defense theory. Prussia provoked three wars within a decade (in 1864, 1866, and 1870) at precisely the time when its leaders believed railroads strongly favored defense. Whereas some European observers viewed the stunningly quick and decisive German victories in these wars as evidence of the offensive advantage of railroads, the dominant view within Germany in the period leading up to World War I was that the spread of railroads would make offensive operations more difficult. Nevertheless, and even though Germany seemingly had the most security to gain from defensive improvements, German leaders relentlessly strove to utilize railroads for decisive offensive warfare. Contrary to offense-defense predictions, the perceived defensive advantage provided by railroads did little to ameliorate Prussia's—and then Germany's—aggressive ambitions.

Military Outcomes

What impact did railroads have on the objective offense-defense balance? As discussed in the previous chapter, any assessment of a given technology's offensive or defensive impact on combat outcomes must control for differences in power and skill. (Power and skill should be taken to mean the relative number and quality of the opposing forces, as well as the relative wisdom with which these forces and the given technology are employed in battle.) The most useful historical cases to examine for present purposes would thus be conflicts between adversaries with roughly evenly matched forces employing railroads in a reasonably wise and competent fashion. Of course, few wars in history have pitted equally strong adversaries across all the relevant variables that underpin military capability. All things being equal, however, cases that better approximate these conditions will be more suitable for testing offense-defense theory.

The advent of railroads coincided with several relatively short and decisive conflicts in Europe between 1850 and 1871. This would appear to support the hypothesis linking mobility improvements and offensive advantages. A closer look at the military campaigns of this period, however, shows that the quick and decisive battlefield outcomes resulted primarily from large asymmetries in military power and skill rather than from any inherent offensive advantage of railways. The absence of war among the European great powers between 1871 and 1914 makes an overall judgment of the offense-defense impact of railroads difficult, but evidence from the American Civil War just before this period and World War I just after suggests that railroads, if anything, favored the strategic defender.

Early Conflicts

Two early demonstrations of the military utility of railroads occurred in the 1850s: during an escalating crisis between Austria and Prussia in 1850 and in the war between Austria and France in 1859. The 1850 conflict occurred in the wake of the Prussian king's proposal to unify all German territories under a "Prussian Union," thereby marginalizing the authority of the Austrian Hapsburg dynasty. Beginning in October, as relations between the two powers deteriorated, over the course of twenty-six days the Austrians mobilized and transported seventy-five thousand soldiers (and eight thousand horses and one thousand wagons) by rail from Hungary and Vienna to Bohemia, where they threatened Prussian territory. Prussia responded with its own mobilization in November. Austria faced some difficulties in moving its forces via the railroad, but Prussia completely bungled its own mobilization. As one historian describes the chaotic scene, "Men, animals, and supplies piled up at loading centers and shuttled aimlessly from station to station on trains whose destination was a mystery."[2] The Prussian debacle stemmed in part from the fact that its army had not been mobilized in thirty-five years, but the most important factors centered on the railroads. At the time, Prussia had no military officers or civilian officials with railroad mobilization experience, possessed only a small rail network, and lacked enough locomotives and rolling stock to deploy sufficient forces. Facing almost certain defeat in the event of war, Prussia agreed to demobilize its army and capitulate to Austrian demands to drop plans for a Prussian Union.[3]

In 1859, France supported the Italian state of Piedmont in its effort to expel Austria from northern Italy. The resulting war was the first to see the large-scale use of railroads for strategic concentration, operational reinforcement, and even tactical movement of troops. Austria, improving marginally on its railroad mobilization performance in 1850, was able to assemble roughly 150,000 troops in twenty days for the invasion of Piedmont. This time, however, Austria's utilization of the railroads was far inferior to that of its adversary. France mobilized and deployed about the same number of troops by rail to the theater in eleven days—a rate of transit almost twice as fast as that achieved by the Austrians—and continued to shuttle troops, horses, and supplies to the theater efficiently over the course of the three-month campaign. Once concentrated in the theater, French commanders (under the direction of Napoleon III) used railroads effectively for reinforcement and tactical movements, whereas Austrian movements were tentative and poorly managed. As the top Austrian general commented, "The enemy soon displayed a superior force, which was continually increased by arrivals from the railway." It is difficult to assess the impact of railroads on the offense-defense balance in this case, not only because of the important disparity between the adversaries in skill at employing railroads but also because of other more important differences in combat power and ability. On the one hand, the French and Austrian forces that clashed in the decisive battles of Magenta and Solferino in June 1859 were equal in number (each side fielding about 60,000 and 160,000 troops, respectively). On the other hand, although both sides suffered from blundering leadership before and during the battles, France prevailed because its commanders were slightly less inept and its troops were more experienced and better equipped and had higher morale than the multiethnic forces of the Austrian Empire. It seems likely that Austria would have been defeated by superior French forces even in the absence of railroads.[4]

Both of these cases from the 1850s foreshadow the tremendous military importance of railroad mobilization and transportation. However, because of large disparities in skill between the main adversaries, neither case can shed much light on the effect railroads had on the offense-defense balance.

American Civil War

The American Civil War (1861–65) was the first war in which railroads played a crucial and widespread role in military operations throughout the conflict. In 1860, the United States had thirty thousand miles of rail, giving it a larger rail network than the rest of the world combined.[5] The unprecedented size of the Union and Confederate armies and the vast geography of the United States required that railroads be used extensively for troop transport and supply. Railroads were put to use at virtually every level of warfare—strategic, operational, tactical, and logistical—and the control of railroad lines frequently became an important military objective in itself.

Did railroads better aid the defender or attacker in the Civil War? Offense-defense theory would expect that increased mobility provided by the railroads would favor the attacker. Indeed, both armies frequently used railroads to concentrate forces for attack or counterattack, as well as to supply ongoing offensives into enemy territory. Just as often, however, railroads were relied on to help hold off enemy offensives. Several of the most dramatic instances of offensive and defensive operations are discussed below. On balance the evidence indicates that railroad mobility did not shift the offense-defense balance toward offense. To the extent any assessment can be made, the better argument is that railroads favored defense by making it harder to fight quick and decisive campaigns and thus helping to prolong the Civil War.

The war pitted unevenly matched adversaries. Available manpower favored the North by a ratio of five to two. The Northern states of New York and Pennsylvania each had more industry than the entire South, with the South's share of national industrial output in 1860 at just 7.5 percent. Although the South was primarily an agricultural entity, it fell far short of the North's production of foodstuffs (wheat, corn, oats, livestock), which were more important than cotton production for a prolonged war.[6] The Union victory in the Civil War was due in large part to these disparities in power, but a crucial factor in the Confederate defeat was the huge inequality between the Northern and Southern railroads.

The South's deficiency in overall material power and capability was especially acute in railroad infrastructure, resources, and management.[7] At the start of the conflict the Confederacy had almost nine thousand miles of railroad, and the Union had over twenty-one thousand miles. Beyond the great disparity in overall track mileage, the Confederate network was inferior on several dimensions. First, the South faced serious disadvantages in terms of its railroad infrastructure. For a variety of geographic, demographic, political, and commercial reasons unique to Southern states, the main railroad construction phase in the 1850s had resulted in a seriously fragmented and disjointed rail network. Southern railroads mostly ran along north-south lines (which hindered the ability to shift forces along the front running east to west and between the eastern and western theaters of the war); lacked compatible rolling stock and track gauges; and had many vital gaps between lines, even when different lines served the same city.[8] For example, in 1861 five railroad lines served the Confederate capital of Richmond, Virginia, but they did not connect with one another. The result was that passengers and freight had to be unloaded at one station, hauled across the city by wagon, and reloaded at another station. More than a dozen other major Southern cities faced the same problem.[9] To make matters worse, the best lines in the Confederacy were in strategically vulnerable or inconvenient locations (for example, Virginia and Tennessee in the former category, Florida in the latter).

Second, the resource constraints faced by the South in all aspects of the war effort were especially severe in terms of manufactured railroad components and supplies. Of the 470 locomotives built in the United States in 1860, only 19 were produced in the South. Moreover, much of the South's existing track had been cheaply built for light traffic. The Confederates thus faced an initial and constant dearth of locomotives, cars, and rails. Because of competition for inadequate stocks of iron and labor, not a single new rail line was produced in the South after 1861, and only four hundred miles of additional track was laid during the war. By comparison, the Union added four thousand miles to its network and devoted enormous resources to protecting, maintaining, and expanding its railroad capabilities during the war.[10]

Third, the Confederate leadership failed to effectively manage and mobilize its railroad resources. Both the Union and Confederate governments faced the same challenge of securing cooperation from privately owned railroad lines for the war effort: standard rate agreements for moving troops and supplies needed to be negotiated, lines needed to be better integrated, schedules needed to be coordinated, and the competing demands between the war effort and railroad industry for labor and scarce resources needed to be managed. In early 1862, the Union Congress gave the president unprecedented power to regulate the operations and take possession of any private railroads. The secretary of war subsequently created the United States Military Rail Roads (USMRR) to supervise and direct rail transportation. Although the USMRR was given authority over all U.S. railroads, the government rarely needed to exercise its sweeping regulatory power in the North (where the shadow of that authority helped generate commercial railroad cooperation throughout the war) and instead focused on acquiring, building, and maintaining rail lines in occupied Southern territory. By contrast, the Confederate government and army never made a serious effort to coordinate, regulate, or control the South's railroads. Private railroads routinely gave priority to shareholder interests over military interests—for example, by rejecting standardized shipping rates for military supplies and refusing to give transport priority to government freight over private goods—and generally frustrated government attempts to better organize and integrate the Southern railroad network. The lack of centralized control stemmed from the adamant commitment to laissez-faire and states' rights found throughout the Confederacy, but it had an enormously detrimental effect on the Confederate war cause. The Southern rail network progressively deteriorated and the Confederacy's ability to supply and reinforce troops precipitously worsened as the war went on. It was not until February 1865, when the outcome of the war was a foregone conclusion, that the Confederacy finally passed legislation giving the government authority to take control of any railroad out of military necessity.[11] In short, the Union held an advantage over the Confederacy in every facet of the railroads, and the Confederacy compounded its problems with massive mismanagement. These deficiencies contributed greatly to the ultimate Confederate defeat.

Despite the gross disparity in power between the North and the South in the Civil War, an assessment of the offense-defense impact of railroads might still be feasible. The war saw many dramatic instances of forces being moved and concentrated by rail for both offensive and defensive purposes. These operations shed some light on the issue.

The Union fought much of the war on the strategic offensive. Railroads contributed greatly to the overall effort, especially as the Union army shifted from an initial strategy of slow encirclement and strangulation (the "Anaconda Plan") to relentless penetration of Southern territory in the last years of the war. Railroads allowed Union forces to expand their size, move faster and farther, and supply themselves from more distant bases. The logistical benefits of the railroad were especially crucial, as the Union army depended on long lines of communication and supply as it drove deep into the South. The best example is General William Tecumseh Sherman's Atlanta campaign through northern Georgia beginning in May 1864. The campaign was conducted with a force of 110,000 men and 35,000 animals, which was supplied entirely by a single railroad line extending almost five hundred miles from Louisville to Atlanta.[12] Sherman later argued that the offensive would have been impossible without the railroad, even though he was constantly worried about the vulnerability of his supply line ("these railroads are the weakest things in war," Sherman complained).[13] It is worth noting, however, that the offensive impact of railroads in this case depended on Sherman's ability to effectively seize, rebuild, and maintain his rail line in the face of almost daily enemy attacks as he advanced deeper into enemy territory. (As discussed below, the Prussians were unable to accomplish this feat in their invasion of Austria and France shortly after the American Civil War.) Moreover, Sherman was able to defend his long and tenuous rail line only because of large advantages in resources, skill, and troop numbers. Railroads were crucial to the campaign but, in Sherman's words, only "because we had the men and means to maintain and defend them in addition to what were necessary to overcome the enemy."[14] Not only did Sherman have enough forces to continue stretching his flanks while defending the rail line against Confederate raiders, but also the USMRR's Construction Corps performed exceptionally, rebuilding eleven bridges, laying seventy-five miles of completely new track, and reconstructing many miles of raided or newly seized destroyed track in the course of the two hundred-day campaign. Whereas the railroads were a crucial factor facilitating Sherman's offensive campaign to Atlanta, the difficulty in maintaining logistical support for his advancing armies led Sherman to cut the railroad umbilical cord for his historic march to the sea in November 1864.[15]

Overall, the benefits railroad mobility offered to invading forces were more than offset by the advantages railroads gave to forces fighting on the strategic defensive. Outnumbered and outgunned, the Confederate army depended on the ability to rapidly concentrate widely separated forces against individual portions of the Union army and to retreat along interior lines whenever necessary.[16] The strategic importance of railroads for the defender was demonstrated at the Civil War's first major battle at Bull Run, fought near Manassas, Virginia, in July 1861, when the Confederates defeated a Union offensive with the timely arrival of reinforcements by rail from the Shenandoah Valley.[17] In early 1862, the fall of two key Confederate fortresses on the Tennessee and Cumberland rivers in the west opened the way for a potentially devastating invasion by the Union army. The Confederates concentrated over forty thousand troops by rail from as far away as Florida and New Orleans to the railroad juncture in Corinth, Mississippi, to block the Union advance. Confederate forces moved north from Corinth for a counteroffensive at Shiloh, Tennessee, in April 1862; the counterattack failed, but the railroads then proved valuable in evacuating troops from the region.[18]

The longest and most famous Confederate troop movement occurred in September 1863 in response to a Union offensive south into Tennessee, an attack that threatened to expose Georgia to invasion and lead to the capture of Atlanta. General Robert E. Lee sent thirteen thousand soldiers under General James Longstreet nearly a thousand miles from Virginia to northern Georgia over the course of twelve days to reinforce the besieged forces of General Braxton Bragg. About half of these troops reached Bragg in time to be decisive in defeating the Union attack at Chickamauga.[19] The Union army quickly reciprocated with its own massive strategic transfer of troops by railroad for defensive purposes. After their defeat at Chickamauga, the shattered Union forces retreated to Chattanooga, Tennessee, where they were put under siege by Confederate forces. In the longest and largest railroad troop movement attempted by the Union, twenty-three thousand reinforcements—along with equipment, horses, wagons, and artillery—were transported twelve hundred miles from Virginia to Chattanooga in eleven days to repel and defeat the Confederate offensive.[20]

These were the most dramatic operations of the Civil War. Other famous mass movements by rail reinforce the conclusion that railroad mobility favored the strategic defender in this conflict and, perhaps, more generally. From March to June 1862 Confederate Generals Lee and Stonewall Jackson grasped the military advantages of railroads as they shifted large numbers of soldiers by train back and forth between the defenses of Richmond and the Shenandoah Valley. In July 1862, the Confederate army undertook its largest single troop movement by rail when thirty thousand men were sent over 770 miles in one week from Tupelo, Mississippi, to Chattanooga (via a roundabout way through Alabama and Georgia in order to avoid Union forces). In August 1862, at the second battle of Bull Run, Confederate forces moved by rail and foot to strike the flank and rear of the Union army, which had cornered Stonewall Jackson.[21]

It is difficult to reach a definitive conclusion about which side in the American Civil War benefited the most from railroads or, at an even more general level, whether railroads on balance favored attackers or defenders. However, the bulk of the evidence suggests that the South was more dependent on the railroads than the North and that railroads, if anything, offered greater advantages to the defender over the attacker. On the one hand, one could argue that railroads allowed attackers, such as the Union forces, to overcome the inherent disadvantages faced by an invading force operating on exterior lines over great distances.[22] Union forces relied heavily on railroad lines of communication to support offensive campaigns into hostile territory. On the other hand, railroad mobility appears to have given defenders operating on interior lines, such as the Confederate forces, an even greatly ability to rapidly reinforce threatened points. In effect, railroad mobility acted as a force-multiplier for the defender, especially a numerically inferior one. Moreover, because attackers relied on existing rail lines in enemy territory to support offensives, defenders were often able to predict likely invasion routes, thus allowing for the preemptive destruction of these lines and concentration of defending forces. If defeated in battle, defending forces could retreat and be resupplied and reinforced by rail more quickly than the attacker could, thus reducing the significance of tactical defeats.[23] Perhaps the strongest evidence from the Civil War supporting the conclusion that railroads gave defenders a comparative advantage was that the top Union generals consistently focused on paralyzing the Confederacy's railroad network, and the Union shifted its overall strategy in 1864 to place even more emphasis on occupying and destroying Confederate railroads in order to bring the war to an end.[24] In short, it appears that railroads prolonged the Civil War by making offensive warfare harder and less decisive.

Wars of German Unification

Prussia's quick and decisive victories in the wars of German unification—against Denmark (1864), Austria (1866), and France (1870–71)—have commonly been attributed to the offensive advantage of railroads. Although the outcome of the Danish War was a foregone conclusion given across-the-board power disparities, Prussia's victories over Austria and especially France were surprising, given that both states appeared to have greater overall military strength and fighting experience as well as extensive rail networks. What else could explain how upstart Prussia could invade and defeat in short order the two recognized great powers of the continent besides the role that railroads played in benefiting the attacker? In fact, the mobility provided by Prussian railroads was less decisive for the outcomes of these wars than superior Prussian strategy, planning, organization, and command. Contrary to offense-defense theory, the battlefield evidence from these wars indicates that railroad mobility bolstered defensive more than offensive military operations.

Danish War   In February 1864, Prussian and Austrian forces attacked Denmark in order to seize the disputed provinces of Schleswig and Holstein. Denmark's hopes for foreign intervention did not materialize, and Danish resistance was crushed in two short campaigns that ended in July. The imbalance of power between the adversaries was enormous: the combined Prussian and Austrian forces dominated Denmark in army organization, military doctrine, troop training, logistics, technology and weaponry, and sheer numbers of deployed forces. For example, in the main battles of July roughly twenty-five thousand Danes faced over fifty thousand Prussians alone. This asymmetry makes any interpretation of the impact of railroad mobility difficult, but the war further indicated the potential military benefits (and even some pitfalls) of railroads. The newly formed Railway Section of the Prussian general staff played an important role in the planning, conduct, and supply of the successful Prussian offensive into Denmark. At the start of the conflict Prussia used railroads to move over fifteen thousand troops and horses to the theater in just six days—a third of the time required for transport by road.[25] The Prussian campaign against Denmark involved only the small-scale use of railroads, however, and offers few lessons for understanding the offense-defense balance.

Austro-Prussian War   Railroads played a larger role in Prussia's quick and decisive defeat of Austria and Austria's allied German states in the summer of 1866. In just three weeks Prussia used five main and several secondary railroad lines to deploy three armies to the Austrian frontier along a 310–mile salient formed by Silesia, Moravia, Bohemia, and Saxony. Seven days after marching into the decisive theater of Bohemia the Prussians completed an outflanking maneuver and routed the main Austrian force at the battle of Königgrätz.[26] The outcome of the Austro-Prussian War was shaped by several factors, including the incompetent Austrian commander-in-chief (General Ludwig Benedek); the technical superiority of the Prussian breech-loading rifle over the Austrian muzzle-loader, as well as the prudent Prussian doctrine of maneuvering infantry to assume the tactical defensive to take advantage of modern firepower (topics that will be addressed in the next chapter); and the organizational, planning, and strategic genius of the Prussian general staff. These crucial asymmetries in power and skill again complicate the matter of determining the offense-defense impact of railroads, but several observations about the role of railroads in this conflict are worth noting.

First, railroad mobility favored the attacker (Prussia) in at least one significant way predicted by offense-defense theory: mobility improvements multiplied the attacker's inherent advantage of the initiative. The strategic mobility provided by railroads allowed Prussia to expeditiously transport over 200,000 soldiers (along with fifty thousand horses and supplies) from the homeland to the theater of operations 150 miles to the south. It is inconceivable that Prussia could have successfully launched such a strategically and geographically ambitious offensive without railroads. Even though Austria was the first state to start arming and possessed a larger army overall, Prussia was able to mobilize and deploy more battle-ready troops to the field more quickly than the Austrians could. (Despite fighting on its home territory, an Austrian force of 240,000 faced a Prussian army of 250,000 at Königgrätz.) Second, the relative benefits of strategic mobility for the defender are hard to discern in the Austro-Prussian War because of the disparity in railroad management and capability. The Austrians had a mobilization plan that was shoddy at best, the transportation of its forces by railroad was unorganized and inefficient, and Austria could use only a single double-tracked railroad line from Vienna to the frontier to concentrate troops. By contrast, the Prussian general staff had in place detailed mobilization procedures and railroad timetables well before the outbreak of hostilities and could make use of a superior railway network. Prussia was thus able to mobilize in half the time it took Austria to do so.[27]

Finally, the contribution of railroads to the attacker's success in the Austro-Prussian War should not be overstated. A closer look at the Prussian offensive reveals that railroads created not only some crucial shortcomings for the attacker but also key advantages for the defender. Strategy aside, the dispersed location of existing Prussian railroad lines meant that Prussian forces *had* to be deployed along an enormous arc of the front, with huge gaps between concentrations. After the initial deployment by rail, Prussian forces set off on foot to close these gaps before the attack. With the repositioning of forces and eventual advance into Bohemia, the Prussian armies quickly outran their supply convoys, leaving food and fodder rotting at congested railheads, and were forced to live off the land. Interestingly, the need to live off the land instead of depending on supplies from railheads gave Prussian forces an unexpected freedom of movement that facilitated its decisive outflanking maneuver. In fact, from the crossing of the Austrian border through the decisive battle of Königgrätz, railways were irrelevant to the Prussian offensive.[28] Other events during the course of the war illustrate that railroad mobility could be a boon for retreating defenders. For example, on two separate occasions, large numbers of Austrian forces were able to retreat by railroad and demolish the tracks as they went; a third force retreated by rail but failed to destroy the tracks, thereby allowing the Prussians to continue their attack.[29] In sum, the historical evidence from the Austro-Prussian War shows clear military advantages in the possession of railroads (at least when effectively employed) but is less decisive on the issue of offensive or defensive advantage. Although the balance of power was closer in 1866 than in 1864, lessons about railroad mobility are clouded by the enormous organizational and planning advantages possessed by Prussia.

Franco-Prussian War   Prussia's decisive defeat of France in the summer of 1870 offers an even starker case of unevenly matched adversaries, especially in terms of the utilization of railroads.[30] The Prussian army—much as it had done four years earlier—used superior strategy, doctrine, and organization to achieve a decisive numerical advantage on the battlefield. Whereas the French could muster four hundred thousand at the height of the conflict, the Prussian army deployed more than one million men.[31] In general terms, the French army had no concrete war plan, an inferior mode of military recruitment, no general staff chief, was led by inept officers and filled with undisciplined troops, and was based on a fundamentally flawed organizational and administrative structure. French war planning, mapping, and war-gaming were virtually nonexistent.[32] Although France had an excellent rail network, French strategists (unlike their Prussian counterparts) failed to understand the military requirements for assembling and transporting a mass army.[33] The utterly incompetent French mobilization and concentration of forces at the outset of the war was the single most important factor leading to France's defeat.[34]

Prussia successfully goaded France into declaring war on July 15, 1870. Both sides began mobilizing their forces on July 16. Prussia's military travel plan, which had been practiced and polished for over two years, worked like clockwork. By August 3, the nineteenth day of mobilization, Prussia and her German allies had used nine railroad trunk lines to concentrate over 450,000 troops on the French frontier. These forces arrived in excellent condition, fully equipped and ready to pursue their orders to "seek the enemy main force, find it and attack it." When Prussia invaded France on August 6, the French had been able to concentrate only about 250,000 men. Moreover, these forces were in no condition to fight, as the disorganized nature of the mobilization had left them with inadequate food, equipment, supplies, plans, and leadership. The French defeat was clear by September 1, when Prussian and other German forces had besieged an entire French army at Metz, forced another to surrender at Sedan, and faced a virtually undefended road to Paris.[35]

Other disparities in military skill and capability compounded France's dismal performance, including the flawed French army command and staff system, lack of reserves, and unsuitable tactical doctrine of massed frontal attacks against the Prussian Krupp steel, rifled, breech-loading artillery. Although France had generally superior weapons—specifically, the Chassepot rifle and the Mitrailleuse machine gun—French soldiers lacked proper training or expertise in using these weapons.

The sharp disparity in military skill was not the only similarity with the previous war. The military contribution of railroads beyond initial mobilization was once again problematic, with the largest shortcomings arising in the supply system. In fact, after the initial deployment of forces, railroads were of limited utility for the Prussian offensive into France. Extended railway lines were jammed with traffic, vulnerable to French attacks, and prone to accidents; railheads could not advance at the pace of the invading troops; and the Prussians were unable to get anywhere near adequate supplies from the railheads to the front. The heavy artillery, ammunition, and forces conveyed by railroads did make possible the siege and bombardment of Paris, but this occurred well after the decisive mobile phase of the campaign was over.[36] In short, although the Prussian army appeared more adept at railroad warfare than the French army, the problems of utilizing railroads for operational maneuver and supply were manifest and may even have exposed the Prussians to defeat by a better-organized and more capable defender.[37]

World War I

In August 1914, Germany sought to crush France in a quick and decisive campaign before shifting forces east to defeat Russia. World War I offers a good case for examining the offense-defense impact of railroad mobility because the opening phase of the war pitted relatively evenly matched adversaries, not just in terms of overall military capability but also on the dimension of railroad infrastructure and mobilization skill. The overall balance of forces between Germany and France on the western front at the outbreak of war was roughly even: despite France's smaller population (about 40 million compared with 65 million), its full mobilization strength was almost comparable (at just under 4 million men) and its total deployed force in the western theater was actually greater (2 million French troops versus 1.7 million German). Although each side could claim military advantages in different categories of aggregate military power, such as in troop training or the quality and quantity of various armaments, these advantages were largely offsetting.[38]

Most important, Germany's crucial railway superiority—which was so vital to its success in the Franco-Prussian war of 1870—was long gone. By 1914, all the European states had built extensive railroad networks and grasped the necessity of detailed organization and planning for employing railroads in wartime. France made great strides in improving the strategic infrastructure of its railroads and achieving efficiency at mobilization and concentration. In 1870, the Germans had nine trunk lines running from the interior of the country to the common border with France; the French had four. By 1914, Germany had thirteen, but France's total had risen to sixteen. Moreover, in the years leading up to the war, a new railway section of the French general staff oversaw and subsidized the construction of two ring lines around Paris and two transverse lines running parallel to the Franco-German border in order to satisfy military requirements for versatility in troop movement. The result was that by 1914 France could mobilize its army as fast as Germany, send its forces as quickly to the frontier, and shift troops laterally along the border depending on the circumstances. Unlike the chaos of 1870, the French mobilization and concentration in 1914 proceeded almost flawlessly. From August 2 to August 18, the French used 11,500 trains to move almost two million men to the front with almost no delays. (Of all the trains used in France's mobilization and concentration, only about twenty were late and the maximum delay was two hours.) Germany's mobilization and concentration was equally superb and efficient, but whereas the Germans moved about four times as many men and horses to the front in 1914 as it did in 1870, the French moved six or seven times as many.[39] In sum, Germany and France went to war in August 1914 on roughly equal footing, thus providing a particularly suitable case for evaluating the offense-defense impact of railroads.[40]

At the outbreak of war, railroads everywhere moved soldiers, weapons, and supplies at an unprecedented pace and scale. However, the enhanced strategic mobility conferred by the railroad did not translate into a quick and decisive victory for the attacker. To the contrary, the defensive advantages of railroad mobility can be seen in two of the crucial battles of 1914, both of which turned on the role of railroads: the French victory on the Marne, which halted the initial German onslaught in the west, and the German victory at Tannenberg, which stymied the first Russian offensive in the east.

The Franco-Russian alliance meant that Germany faced a two-front-war problem in 1914. According to the traditional narrative, Germany's operational solution was to launch a massive flanking movement through Belgium and Luxembourg that would wheel around France's eastern fortresses, envelop Paris, cut the French lines of communication, and roll the discombobulated French forces back up against the Franco-German frontier—all before the slower-moving Russian forces could invade Germany in the east. Recent scholarship has challenged this familiar explanation and suggested that German operations had far less ambitious goals at the start of the conflict. (The debate is discussed later in the chapter.) Regardless, Germany's basic plan was to achieve some sort of quick and decisive victory in the west before shuttling its forces east to face the Russia army. Given the large geographic scale and short window of time before Russia completed its mobilization and concentration, Germany's strategy of rapid, sequential offensives would have been impossible without the speed and flexibility conferred by the railroads. Offense-defense theorists might thus consider railroads a boon to the attacker. However, in military operational terms, the benefits of railroads in this context were primarily defensive. After all, the Confederate army in the American Civil War had frequently employed the same basic defensive maneuver of shifting forces by railroad along interior lines. The difference in 1914 was that the Germans had adapted the defensive operations to fit an offensive strategy.

The way the western campaign unfolded reinforces the point about the relative defensive advantages of railroads. For the main German advance through Belgium, Luxembourg, and northeastern France, which began on August 18, German trains successfully concentrated over fifty divisions, including 320,000 men in Alexander von Kluck's First Army on the spearhead of the right wing.[41] The main problem hindering the German advance after this initial successful deployment was the inability of the railheads to keep up with the advancing German forces, a problem that was anticipated by German planners but made worse by Belgian and French sabotage of railroads, bridges, and tunnels. The Belgian railway network was dense and provided a good link between Germany and France, and the railway section of the German general staff did an excellent job in repairing and managing the seized lines, but railroad logistics could not keep up with invading forces. After leaving the frontier railheads, Kluck's army marched over fourteen miles a day for three weeks and relied almost exclusively on horse-drawn transport for forward supply. (Kluck's army had eighty-four thousand horses, which in turn needed two million pounds of fodder a day.) By September 6, as the German advance ground down and wheeled in front of Paris, the constant marching, fighting, and supply problems had degraded the field strength of the right wing of the German invasion by about half, and Kluck's army was over eighty miles beyond its nearest railhead.[42] Meanwhile, the French exploited the strategic mobility provided by their rail network to redeploy forces from the Franco-German border to the interior of the country to defend against the German turning movement. On August 23, the German right wing of 24.5 divisions faced 17.5 French and British divisions. On September 6, after French forces had been rushed back by train, the same German force now faced 41 allied divisions. The French troops arrived just in time for the battle of the Marne (fought between September 6 and 9), which marked a decisive defeat for Germany, its entire western offensive, and hopes for a quick victory.[43] In short, if the German attack in 1914 was made possible only by the railroads, the same could be said for the French "miracle on the Marne" and defeat of the German offensive.

Russia had also made great improvements in railroad construction in the years prior to 1914. It led the world in the rate of railway construction between 1890 and 1913, although it still trailed far behind its European neighbors in many aspects of railroad power (including the number of trunk lines leading from the interior of the country to the frontier, length of track in relation to size of territory, and skill at organization and planning).[44] Most important, Russian strategic railway building was gathering steam, in large part because of French loans and other assistance. The German military had already estimated in 1912 that Russia had cut in half the amount of time it would need to concentrate on Germany's border in the previous five years.[45] Between 1910 and 1914, in fact, the number of trains that Russia could send to the frontier in the event of mobilization rose from 250 per day to 360, and a target of 560 had been set for 1917.[46]

Indeed, when war came in August 1914, Russia was able to mobilize more quickly than expected and launch an offensive against German forces in East Prussia. Facing a coordinated pincer attack of two Russian armies around the north and south sides of the Masurian lakes, the overall numerically inferior German defenders used railroads to defeat the Russian offensive in piecemeal fashion. Originally deployed in the north of the lakes, the Germans pulled back and marched two corps to the south while simultaneously shifting an entire corps by rail much farther along a circuitous route until it could turn the left flank of the Russian army in the south. This rapid operational concentration allowed Germany to encircle and crush one Russian army at the battle of Tannenberg in late August before turning northeast again to defeat the second Russian army at the battle of Masurian lakes in early September.[47] Although incompetent leadership contributed to the failure of Russia's offensive, railroads once again allowed the defender to react and concentrate more quickly against the attacker.

In sum, the impact of railroads on the opening battles of World War I, as well as on combat outcomes in the previous fifty years, contradicts the mobility-favors-offense hypothesis. Attackers were often slowed by the need to repair damaged rail lines as they advanced, while defenders frequently used their own lines to rapidly retreat, concentrate, and counterattack. Moreover, as the defensive stalemate took hold on the western front in World War I in particular, railroads transported a relentless stream of supplies that allowed the adversaries to sustain huge armies in the field for years. On balance, the greater mobility produced by railroads played a powerful role in denying quick and decisive victories to attackers.

Political Outcomes

Railroads did not appear to confer an advantage on the attacker. Did political and military leaders believe otherwise? If so, did these leaders act according to offense-defense predictions? These are the crucial questions for understanding the broader utility of offense-defense theory as a guide to international politics. After all, the theory is based on the notion that actual offense dominance is relatively rare, whereas perceived offense dominance is a common cause of war.[48] If leaders perceived that railroads were offense dominant, we should find evidence that the spread of railroads made states feel more vulnerable and more inclined to behave aggressively. If leaders perceived defense dominance, the spread of railroads should have made states feel more secure and act less belligerently.

Early Views of Railroads

As commercial railroads spread across the European continent and the United States in the 1830s and 1840s, civilian and military leaders debated the potential strategic utility of this new technology. Much of the early discussion in pamphlets and articles was highly speculative. Some commentators warned that the building of railroads would only facilitate foreign invasion. British observers, for example, feared that railroads might facilitate a sudden French concentration at the English Channel ports and invasion of the island. Similarly, in the earliest days of the railroad era the Prussian army often opposed the construction of new rail lines on the grounds that such lines would help an enemy overrun homeland fortresses.[49]

The dominant view, however, was that railroads would greatly favor the defense by improving the ability to shift troops to counter any threatened sector of a frontier. In the United States, where bitter memories of the British invasion and destruction of the U.S. capitol in the War of 1812 were still fresh, private railroad promoters repeatedly (if predictably) emphasized the potential contribution of railroads to the national defense in seeking governmental aid for their projects. More objectively, the U.S. Congress and Department of War closely examined the military potential of railroads from the beginning. In 1836, the secretary of war concluded in a report to Congress that the new mobility provided by railroads could neutralize the great advantage possessed by a seaborne invading force, such as the British demonstrated in 1813 and 1814, in which the attacker could choose which seaport to attack and destroy before defensive reinforcements could arrive. In 1837, a House committee agreed that even numerically superior forces, which the United States had at its disposal for defending eastern cities against coastal invasion in the War of 1812, were useless if those forces could not be concentrated quickly enough to defend any single city.[50] Although much of the debate within the U.S. government at the time focused on the extent to which railroads had superseded or diminished the necessity of coastal fortifications, few disagreed with the assessment of an American general's report to Congress that railroads would "render a nation in the position which we occupy, at least ten times more formidable in a war of self-defense, than in an offensive war, against nations of equal numerical strength."[51] Similarly, a number of European commentators, most notably the German economist Friedrich List, advocated building railroads for strategic defensive purposes. List argued that railroads would make it especially easy to defend a centrally located territory like Germany because troops could more quickly reach the frontiers or use the weblike structure of the rail network to encircle advancing enemy forces. "The most beautiful thing about it all," wrote List, "is that all these advantages will benefit the defender almost exclusively, so that it will become ten times easier to operate defensively, and ten times as difficult to operate offensively, than previously."[52] Establishing his credentials as one of the earliest offense-defense theorists, List concluded that because of this defensive superiority the railroad would be the instrument to destroy war itself and bring perpetual peace to the European continent.[53] This sentiment about railroads was echoed by a member of the French Chamber of Deputies in 1833: "If a country could thus speedily carry considerable masses of troops to any given point on its frontiers, would it not become invincible, and would it not, also, be in a position to effect great economies in its military expenditure?"[54]

Other early observers were less sanguine about the political implications of the emergence of railroads, even when they shared the dominant view that railroads conferred a decisive advantage on the defense.[55] Moreover, the perception of defensive advantage did not preclude serious investigation into the offensive possibilities made possible by the railroads. For example, the Saxon captain Karl Pönitz speculated about a German invasion of France that might be facilitated by a rapid concentration of forces by rail on the west bank of the Rhine, sudden thrust into French territory, and further advance along captured French tracks.[56] Nevertheless, by 1850, it was widely accepted that railroads gave an overwhelming advantage to defensive military operations.

American Views of Railroads before the Civil War

There is little need to discuss the role of strategic perceptions of railroads as a cause of the American Civil War. No offense-defense proponent argues that the outbreak of the war—in terms of either a deep or a proximate cause—was in any way conditioned by perceptions of the offensive or defensive nature of railroads. Indeed, few military thinkers in the North or South had given much consideration to whether railroads would play an important military role in the coming conflict. Several Union generals, including Grant and Sherman, closely followed the war between France and Austria in Italy in 1859, in which railroads were used extensively, and Sherman was clearly aware of the impact of railroads on moving and supplying troops.[57] It was not until after the war started, however, that American military commanders formed more concrete views on the military advantages of railroads based on their employment by both sides in offensive and defensive operations. As discussed earlier in this chapter, one of the definitive conclusions that can be reached is that Grant and Sherman clearly appreciated the contribution of railroads to the Confederate defense strategy. Both generals devoted much effort to disrupting and destroying the Confederate railroad network and rail centers as a means of breaking the South's ability to move and supply forces along interior lines.

Prussian Views of Railroads during the Wars of German Unification

Prussia at mid-century was vulnerable. It was surrounded by strong potential enemies (France, Austria, and Russia by land, Britain by sea) without any significant natural barriers. Whereas the Prussian army was inexperienced in war—not having fought since 1815—other European armies had been busy fighting at least small wars since that time and, most recently, against each other in the Crimean War (1853–56). Given this precarious position, Prussian leaders were especially quick to recognize and embrace the defensive advantages of railroad mobility. Yet the essentially accurate perception of the strategic implications of railroads did nothing to discourage Prussian aggression. Contrary to offense-defense theory expectations, Prussian leaders opportunistically capitalized on the defensive strengths of railroads in order to pursue an expansionist foreign policy.

A thorough discussion of Prussian politics and diplomacy during the period of German unification is unnecessary for present purposes. Few would disagree that all of the key players in Prussia—Otto von Bismarck, King Wilhelm I, and Helmuth von Moltke—were bent on expanding Prussian power in Europe by unifying the German states under Prussian control. From the day Bismarck was appointed prime minister in 1862 he openly proclaimed that this objective would probably have to be achieved through war.[58] Although technically France declared war on Prussia in 1870, Prussia is correctly seen as the aggressor in all three wars of the decade. The crucial question for offense-defense theory is whether perceptions of defense dominance in any way diminished or discouraged Prussia's political or military ambitions. They did not.

The evolution of Prussian strategic thought on railroads is inseparable from Helmuth von Moltke, chief of the general staff from 1857 to 1891. Moltke's predecessor since 1848, Carl von Reyher, shared the contemporary view of railroads as a defensive innovation and sought to solidify this advantage. In particular, Reyher warned that new lines being constructed in eastern Prussia should have narrower gauges than the Russian tracks so as to preserve the ability to support threatened frontiers without abetting an invading force. Moltke's thinking on railroads was more advanced. He had taken an interest in the military possibilities of railroads as soon as they appeared on the continent. His attention was first drawn to the importance of the new form of transportation through an analysis of the Russian-Turkish campaign of 1828–29, which Moltke wrote in the early 1840s; he read the work of Friedrich List on the strategic potential of railroads and in 1841 was invited to join the board of directors for a new Berlin-Hamburg line.[59] As commercial railways spread across the land—by 1850, Germany had over three thousand miles of track—the Prussian army experimented and learned. In 1839, eight thousand troops were moved from Berlin to Potsdam during a regular maneuver. In 1846, twelve thousand troops were sent by rail to the Silesian frontier when disorder broke out in Cracow. In 1850, as chief of staff of the IV Corps, Moltke experimented with railroads in corps maneuvers.[60] The bungled mobilization of November 1850, in which it took two months for Prussia to mobilize four hundred thousand troops, was analyzed closely by Moltke. He understood that large-scale mobilization would require a much more extensive railway network, greater military control of lines and facilities before and during conflict, and—above all—far better organization and planning. The Prussian general staff sent a team of observers to the Crimean peninsula in 1854–56, and they returned with the lesson that railroads were crucial for supplying armies in combat. Soon after his appointment as chief, Moltke's general staff issued a draft statement on the use of Prussian railroads for troop movement and, in September 1858, Moltke tested these ideas in official maneuvers. The exercises again revealed the enormous military potential of railroads, as well as the need for greater planning and coordination.[61]

Moltke and his staff paid close attention to the use of railroads in the 1859 war between France and Austria in northern Italy. With an eye west toward France, Prussia herself had mobilized during the conflict, but, as in 1850, the process was a disaster. In this case, one of the major problems concerned civilian-military coordination of rail movement. Immediately after the 1859 war, Moltke thus proposed that the Prussian railroads be more explicitly organized for war. Moltke also learned a number of valuable lessons from a subsequent study of France's railroad organization and capabilities. In an 1861 essay, "Importance of Railroads," Moltke suggested that railroads conferred greater benefits on the defender: "Basically, each and every addition to communications, especially to the railroads, must be considered a military advantage. All railroads in the rear of an army are useful while it is stationary, as well as during a retreat. During an advance the railroad must first be captured. This means that the railroad, sure to be destroyed by the enemy, has to be repaired. The country on both sides must be under control before one can use it."[62]

The defensive advantages of railroads were also crystallized for Moltke by his understanding of the American Civil War. Some writers have claimed that Moltke and other military observers in Prussia and Europe ignored the nature of fighting and operations in the Civil War. This claim is incorrect and largely based on Moltke's alleged remark that the Civil War was nothing but a matter of "two armed mobs chasing each other around the country, from which nothing could be learned."[63] The argument is taken up in the next chapter, as the debate has largely centered on the issue of whether Europeans understood the impact of new firepower innovations, but the fact of the matter is that Moltke was kept well informed about the Civil War from close colleagues, from an observer he personally sent to the conflict, and from the published impressions of Prussian observers.[64] Moltke thus correctly recognized that the Civil War was the first great railroad war. Prussian studies of the American Civil War provided the impetus to create a railway section in the general staff in 1864 and a field railway section in 1866, with the latter modeled directly on the Union's USMRR Construction Corps.[65] Most important, given the similarities between German and American topography—with north-south running rivers as natural defensive barriers—Moltke saw the advantages of expanding Prussian railroad lines running east and west in order to complement the natural defenses.[66]

All of this practical knowledge of railroads in war pointed to one conclusion: railroads were less useful for invading forces but of great use to the defender for mobilizing, concentrating, and supplying forces. Moltke concluded that railroads provided a strong boost to Prussian defense be-cause—if an extensive enough railroad network existed and with proper planning and coordination—troops could be deployed or redeployed rapidly to meet invaders at the frontier. Prussian defenders would then have full use of their own railroads in retreat, while advancing attackers would not. In short, Moltke concluded that only the defensive mobility provided by railroads could counterbalance the larger forces of Prussia's neighbors.[67] Time and mobility were essentially the key—time to mobilize, transport, and concentrate against Prussia's enemies before they had a chance to push the attack—and time and mobility could best be bought through the efficient planning and use of railroads. These views formed the basis for Moltke's famous dictum "Build no more fortresses, build railways."

The conclusion that railroads benefited defenders was a happy one for Moltke and his staff because the dominant expectation in the early 1860s was that Prussia would soon be attacked by its neighbors and would have to fight on the defensive from the outset of any war. Moltke's first official war plans against France and Austria were defensive to the core. Both plans aimed at using railroads to deploy forces to areas where they could hit the flanks of the invading force.

The perceived benefits of an improved defense conferred by railroads did not, however, diminish Prussia's ambitions. Moltke firmly believed that offense was the more desirable form of warfare because only strategic offensive operations could achieve the fundamental goal of war, which was the destruction of the enemy's forces. Moltke's desire for rapid offensive success formed the basis for a new Prusso-German school of warfare as practiced through World War II, but it originated in the practical security realities of the moment—i.e., Prussia wanted to avoid fighting wars of attrition against numerically superior enemies.[68] In this context one can better understand how Moltke channeled the defensive benefits of railroads to offensive purposes. His original war plan against France was defensive, but in January 1870, six months before the war, Moltke appended his "Importance of Railroads" memo with the following passage: "The enormous influence of railroads on the conduct of war has unmistakably emerged in the campaigns of the last decade. They enormously increase mobility, one of the most important elements in war, and cause distances to disappear. . . . [R]ailroads serve the defense better than offense. . . . The main point [is] to strengthen the offense with the means of the defense. This is preferably done by the new element—the railroads."[69] In short, as the following discussion of the lead-up to each war shows, the evolution of Prussian strategy during the period of German unification offers a clear rebuttal of offense-defense theory.

Danish War   Bismarck wanted to annex the territories of Schleswig and Holstein, which were legally and dynastically tied to Denmark. Days after taking office in September 1862, Bismarck asked Moltke for an appraisal of war against Denmark. Moltke's December 1862 memorandum concluded that Prussia could achieve numerical superiority and thus victory with about sixty thousand troops. There was little doubt that Prussia (and Austria) could muster the necessary forces. The more immediate strategic consideration for Bismarck and Moltke was whether Denmark could be defeated quickly enough to preclude major power intervention. Moltke predicted that the war might be won quickly if railroads could rush troops into the territories to establish strong defensive positions ahead of Danish forces, which would be arriving by ship and foot (in wintertime).[70]

The use of railroads in the Danish War was small-scale, but Moltke derived several important lessons. One quarter of the Prussian force, about fifteen thousand men, had been moved together by rail to the theater at the outset of the war. Moltke was troubled by the rampant transportation and supply delays involved with even this relatively small operation. It was clear that massed troops could not move nimbly or rapidly. In essays from 1864 and 1865, Moltke argued that the growing size of armies not only necessitated rigorous strategic planning but also a fundamentally different operational approach that linked the initial mobilization, deployment, and opening moves of a campaign. The basic idea was to divide the army into separate mobilizable units, deploy these forces rapidly to different sections of the frontier, have each military unit proceed independently to the battlefield, and not concentrate the whole army until immediately before a decisive battle. Each unit would assemble at a designated rail station, follow precise railroad timetables, and arrive at frontier railheads as a fully functioning force. From that point on the movement of forces into battle would depend on the circumstances. Above all, Moltke's new operational doctrine—to deploy armies separately to minimize logistical friction and to concentrate only on the battlefield ("march divided, fight united")—was aimed at bringing superior numbers to the decisive final engagement with the enemy.[71]

Moltke's operational scheme relied heavily on railroad mobility.[72] This should not be taken as evidence of his belief that offense had become relatively easier. Not only did Moltke still view railroads as favoring the defender, but he also understood that the immediate strategic objective remained defensive: given Prussia's lack of geographic depth, overall numerical inferiority, and likelihood of being attacked, only superior railways and mobilization planning would allow Prussia to overcome these disadvantages and defend itself. Prussian ambitions grew larger—i.e., more aggressive—only when it realized that its attention to the military importance of railroads had not been equally shared by its adversaries.

Austro-Prussian War   Even before the Danish campaign, Bismarck was determined to exclude Austria from a united German state and sought to provoke a war through aggressive diplomacy. In a letter to a Prussian diplomat in Vienna in 1856, Bismarck wrote:

> Because of the policy of Vienna [the Congress of Vienna, 1815], Germany is clearly too small for us both [Prussia and Austria]; as long as an honorable arrangement concerning the influence of each in Germany cannot be concluded and carried out, we will both plough the same disputed acre, and Austria will remain the only state to whom we can permanently lose or from whom we can permanently gain. . . . I wish only to express my conviction that, in the not too distant future, we shall have to fight for our existence against Austria and that it is not within our power to avoid that, since the course of events in Germany has no other solution.[73]

For his part, Moltke regretted that the struggle for political control of Germany would likely bring war with Austria, but he supported Bismarck in provoking the conflict.

Both Bismarck and Moltke expected a hard fight against Austria. The plan from the start was to fight a defensive war. For one thing, an invasion of Austria seemed out of the question because of enormous geographic disadvantages. The direct route to Prague and onward to Vienna was well protected by the Bohemian mountains and strong fortresses. More pressing was the fact that Moltke expected the Austrians to invade first. Specifically, Moltke had to contend with two possible invasion routes for Austrian forces: either the Austrians could drive directly from the southeast into the Prussian territory of Silesia or they could deploy through the German state of Saxony (which was hostile to Prussia and assumed to be allied with Austria) and be poised to march on Berlin, thus threatening the Prussian capital and splitting Prussian territory in two. Moltke was convinced that the main Austrian drive would come through Saxony toward Berlin and that the Austrian invasion force would have superior numbers. The proposed solution was a reliance on railroads to skillfully use space and movement to carry out a series of flanking attacks on the advancing Austrian forces—not from static defensive positions but from wherever Prussian forces could best concentrate and maneuver. Moltke realized that Prussia could improve its defensive striking capability by occupying Dresden in Saxony at the outset of war, given its importance as a railway junction. All of this could be done only if Prussia seized the initiative and declared war as soon as her enemies began to mobilize.[74]

As relations deteriorated, Moltke further developed a plan of aggressive defense against likely Austrian attack routes, using railroads to concentrate forces as quickly as possible for operational flank attacks. However, as memoranda in the winter of 1865–66 and spring of 1866 show, Moltke was increasingly concerned that the war plan was not sufficient to overcome Prussia's numerical inferiority. The basic problem was that even if Prussia could turn back an initial Austrian attack, the unfavorable balance of power—with most of Prussia's neighbors, including the large states of southern Germany, likely to side with Austria—meant that Prussia would almost certainly lose a more protracted war. Moltke thus was attracted to fighting preemptively and offensively. If war was inevitable, as Moltke believed, it might make more sense to attack before Austria and her potential allies were ready. He calculated that Prussia's more extensive railroad network would allow it to complete the concentration of three hundred thousand Prussian troops on the Austrian and Saxon frontier in half the time it would take Austria and her allies to mobilize an equivalent force. As Moltke wrote in an April memorandum to the King, there would be a two- to three-week window when Prussia would have clear numerical superiority; a window when Prussia should fight the decisive battle.[75]

As it happened, Austria's "pre-positioning" of troops on the border and subsequent partial mobilization, combined with King Wilhelm's hesitation at launching full Prussian mobilization, resulted in only a slight Prussian numerical superiority (254,000 vs. 245,000) in the decisive theater.[76] Moreover, for various domestic and foreign policy reasons, Moltke was forced to disembark the Prussian army at railheads farther apart than he would have liked. The result was that Moltke fully abandoned the defensive option—waiting for an Austrian attack before falling on its flanks and rear—and instead pleaded for a prompt offensive concentration into Austria as the only way to protect the dispersed Prussian army.[77] On June 23, 1866, the first Prussian forces crossed the Austrian border. For all practical purposes, the war was over with Austria's crushing defeat at the battle of Königgrätz on July 3–4.

Franco-Prussian War   Offense-defense theory does not provide much leverage for understanding the fundamental political causes of the Franco-Prussian War. In the simplest terms, Bismarck and Moltke desired war against France as the final step toward German unification. Moltke spent years preparing military plans for such a war, and Bismarck successfully provoked France into declaring war on Prussia in July 1870. Offense-defense proponents might be on firmer ground in claiming that Prussian perceptions of railroads contributed to the outbreak, or proximate cause, of the war. The argument would be as follows. Although Moltke's initial war plans against France called for defensive operations, after the Austro-Prussian war, he and his staff realized that Prussia could successfully conduct offensive warfare against France by taking advantage of the strategic benefits of railroads. The Prussians then provoked and won the war against France. In short, according to this argument, perceptions that railroad mobility shifted the offense-defense balance in favor of the attacker made Prussia more inclined to initiate war with its neighbor.

There is no doubt that Prussian leaders believed railroads would help Prussia defeat France and achieve its long-standing political ambition of territorial expansion and control of the entire German confederation. But two major points undermine offense-defense explanations. First, Prussian leaders did not believe railroads were offense-dominant. They continued to believe that railroads, on balance, favored the defender. What they did perceive was that Prussia was capable of mobilizing and deploying a far larger force by rail to the frontier much faster than France. In other words, Prussian aggression was fueled by perceptions of military superiority, not offense dominance. Indeed, as stunning to continental observers as it was at the time, Prussia's quick and decisive victory against France was the logical result of far superior Prussian planning, organization, professionalism, and command.

Second, as the writings of Moltke and other general staff officers show, the Prussian view that railroads conferred an advantage on defenders did not diminish Prussia's aggressive intentions. Instead, the perceived defense dominance of railroads bolstered an offensive military strategy. The paradox arose from Prussia's fundamental strategic dilemma of a two-front war. In the event Prussia went to war with France, Austria and Denmark would likely seize the opportunity to attack Prussia to reverse the losses incurred in their recent defeats. Moltke originally concluded that this dilemma, combined with his belief in the superiority of the French rail network, required that Prussia fight on the defensive at the outset of war.[78] By 1870, however, Moltke's war plan against France had become decidedly offensive. What changed? Moltke had not suddenly come to recognize railroads as offense-dominant. In his January 1870 revisions to the "Importance of Railroads" memo, Moltke wrote that railroads "enormously increase mobility" but also that "railroads serve the defense better than the offense." The shift in the Prussian war plan against France was driven by Moltke's greater appreciation of how the improved strength of defensive warfare in the age of railroads could bolster an attacker facing a two-front-war problem. In Moltke's words:

> The political situation plainly indicates that in the near future, and probably also for a long time to come, we will have either no war or one against neighbors on two sides. In the latter case, we would stand on interior lines of operations between a completely ready and a slowly arming opponent [i.e., France and Austria]. Everything will depend on a correct and timely decision that we use a substantial majority of our active fighting forces first against one and then against the other opponent. By this we put them to double use. However, we need a highly developed railroad net for that purpose and we may say that each and every new construction of railroads is a military advantage.

In the next six months leading up to the war in July, Prussia accelerated railroad construction, practiced and improved railroad mobilization and deployment schedules, and fine-tuned its plans for attacking France.

In sum, according to offense-defense theory, if Prussian leaders perceived that railroads were offense-dominant, we should find evidence that the spread of railroads made Prussia feel both more vulnerable and more inclined to behave aggressively. If Prussian leaders perceived defense dominance, the spread of railroads should have made Prussia feel more secure and act less belligerently. Neither expectation is supported in this case. The perception that railroads favored defenders did nothing to discourage Prussian belligerence in the Wars of German Unification. Instead, Prussia was able to act aggressively through its understanding that because railroads made the defense of Prussian territory easier it would be easier to fight sequential or even roughly simultaneous wars against its neighbors. Military leaders understood correctly that railroads offered greater advantages for the defender than the attacker, but this had no dampening effect on their willingness to wage war.

German Views of Railroads before World War I

In the wake of German unification, all of the European continental powers strove to duplicate German institutions for employing railroads in war. Offense-defense scholars argue that European statesmen believed that railroads favored attackers and that any future war would be short and decisive. According to offense-defense proponents, these misperceptions were a significant cause of World War I. George Quester writes, "The world after 1872 thus theorized that transportation technology now dictated taking the offensive in land warfare, and taking it as quickly as possible . . . [and] continually expected land wars to be short and decisive."[79] For Stephen Van Evera, "had the actual power of the defense been recognized . . . the Austro-Serbian conflict of 1914 would have been a minor and soon forgotten disturbance on the periphery of European politics."[80] Instead, as Quester argues, World War I was "launched on the illusion of offensive advantage."[81] World War I is seen as the paradigmatic case for offense-defense theory, as it appears to show the severe instability and tragic consequences that can result from perceived offense dominance. The offense-defense argument is straightforward and familiar: the mistaken belief that attackers had a great advantage generated tremendous pressure for states to mobilize their armies quickly and strike preemptively. This pressure led to the creation of highly inflexible and interlocking railroad mobilization timetables and offensive war plans that limited the time and space for crisis diplomacy. These military-technical considerations dominated policy and decision making and directly led to a war that might otherwise have been avoided if leaders had appreciated the true nature of the offense-defense balance.

The "cult of the offensive" explanation has previously been challenged by political scientists and historians, but recent scholarship has gone further to discredit the thesis that World War I resulted from such a tragedy of errors and misunderstandings. In particular, the view that rigid, interlocking railroad mobilization plans triggered an unintended spiral of escalation that none of the major powers could reasonably stop—an interpretation made famous by historians A. J. P. Taylor (in his book *War by Time–Table* and other works) and Barbara Tuchman (*The Guns of August*)[82]—has been undermined and replaced by a general consensus about German intentionality and culpability in provoking a preventive war against Russia and France.[83] The remainder of this chapter briefly discusses the role of railroads in German strategic thought before 1914 in order to dispel the notion that a belief in the offensive advantage of railroads led Germany down a slippery slope toward war. The next chapter challenges the broader claim that German leaders believed a coming war would be quick and decisive.

Neither Moltke nor Bismarck came away from the wars of German unification believing that the spread of railroads had commenced a new era of swift and decisive warfare. The general staff was well aware that its rapid battlefield successes in the opening phase of the war against France in 1870 were due to German superiority in mobilizing and concentrating forces by rail, but two ominous conclusions undercut military optimism. First, Germany recognized serious operational limitations in using railroads to carry on the offensive beyond the frontier to eliminate protracted French resistance in 1871. In other words, German leaders understood that a properly conducted railroad mobilization could lead to huge initial gains in battle (especially if the adversary's use of railroads left much to be desired), but it could not ensure eventual victory in war. Second, the Germans foresaw a new phase of intensive railroad construction and organizational planning in France and Russia that would eventually level the strategic playing field. To be sure, German planners understood the enormous advantage that would accrue to the side that more quickly mobilized and concentrated its massive conscript army by rail—thus their mobilization timetables were continuously revised and shortened—but they also anticipated that Germany's adversaries would inevitably seek to emulate this key component behind Germany's recent victories. Thus, Moltke believed that if France attacked to regain Alsace-Lorraine (territory it lost in 1871), Germany should fight on the defensive until it had the opportunity to launch a counterattack. If Germany was forced to fight a two-front war, his inclination was for it to defend in the west against France and launch an offensive in the east against Russia because the latter would still have an inferior railroad network. The important point is that Moltke realized that Germany's numerical inferiority and diminishing railway superiority made a repeat of the stunning success of 1870, much less a repeat of that victory on two fronts, highly improbable.[84] In his last speech to parliament in 1890, Moltke cautioned that the next European conflict would be another "Seven or Thirty Years' War."[85]

The evolution of German war planning and intentions in the decades leading up to World War I is the subject of intense debate. Not only were the war plans of Moltke's successors—Alfred von Waldersee (1888–91), Alfred von Schlieffen (1891–1905), and the elder Moltke's nephew Helmuth von Moltke the Younger (1906–14)—destroyed in a bombing raid in World War II, but it also seems that various influential historical interpretations have been shaped by political and ideological concerns. Terence Zuber, for example, in 1999 sparked a debate that still rages with his thesis that Germany's "Schlieffen Plan," the blueprint for a successful massive encircling operation to knock France out of the war in six weeks before turning against Russia, never existed. According to the traditional account, a master plan for victory was drawn up by Schlieffen in his famous 1905 memorandum, but then modified by the younger Moltke in the years immediately before the war in a way that led to the failure of the campaign and the onset of years of trench warfare. According to Zuber, the 1905 memorandum was nothing more than a political ploy aimed at winning a vast increase in the size of the German army: civilian leaders would hesitate to reject an ambitious war plan that promised victory even if it required a massive mobilization of German manpower and resources. Moreover, Zuber contends, the Schlieffen Plan achieved its place in popular lore only because the postwar general staff, faced with professional extinction, "decided to explain its failure in 1914 by maintaining that it had an infallible plan, which was spoiled by the action of [a few] dead officers," including Moltke, who failed to understand the genius of the original plan that would have brought swift victory.[86]

Some have rejected Zuber's central thesis and others have engaged it in a critical dialogue,[87] but most historians now agree that the best interpretation of old and new evidence is that at no time did German planners settle on a single, tightly coordinated operational plan to annihilate the French army in a massive, wheeling campaign of encirclement deep in French territory. Instead of intending to fight a rigid *offensive* war driven by railroad timetables, both Schlieffen and Moltke understood that Germany's numerical inferiority and two-front-war problem now meant that Germany would have to fight a flexible *defensive* war. Specifically, Germany would allow its enemies to attack and then rely on the mobility provided by its railroad network to launch debilitating counteroffensives. As Hew Strachan notes, "By using interior lines, Germany could switch troops from east to west and vice versa, so transforming its central European position from a liability to an advantage. The benefits were primarily defensive, but they were shoehorned into an offensive framework."[88] The strategic objective would be to use rail mobility to surprise and attack with overwhelming force any vulnerable flank of the invading (or retreating) attacking armies. The basic intent was to decisively win the initial big battles of the war, which would be fought near the frontiers, not deep inside France or Russia. As the French were expected to attack into western Germany via Lorraine or through Luxembourg and southern Belgium, the decisive battles were expected to occur there or, if the French retreated in quick order, in eastern France. Strategic geography thus dictated having a strong German force ready to move into and through Belgium. It is true that Schlieffen and Moltke increasingly focused solely on the western front, realized that Germany's best hope of a decisive flank attack would require a move through Belgium, and strove for shorter and tightly planned mobilization and concentration schedules. It is also correct that Moltke strengthened the left wing of the German armies facing west, which provided grist for postwar claims that he "watered down" the right wing that was meant to deliver the roundhouse Sunday punch to the French army. However, none of these actions fundamentally changed the strategy of seeking to counterattack the open flank of a French attack.[89] Once the war began and German forces fought through Belgium into France, Hew Strachan writes, "Moltke was looking for victory where he could get it, not where some grand design of envelopment suggested."[90] In short, railroads were neither perceived by German leaders as offense-dominant nor incorporated into German plans on the basis of "use it or lose it" logic. Railroads were recognized as a crucial component of modern war in conducting both defensive and offensive operations, but they did not pose the kind of preemptive incentive problems depicted by offense-defense theorists.[91]

In sum, the image of military technology as a runaway train (so to speak) hurtling the major powers of Europe toward a war in 1914 that all wished to avoid does not fit the facts.[92] Kaiser Wilhelm II, Chancellor Theobald von Bethmann Hollweg, War Minister Erich von Falkenhayn, and General von Moltke all believed to varying degrees that a future war involving Germany was inevitable, a preventive war against Russia was desirable, and a broader European war against Russia, France, and Britain was both likely and acceptable as a means to shore up and enhance Germany's power position in Europe. The main objective of German foreign policy during the July crisis was to try to create the appearance that Germany was not the aggressor and to pin responsibility for the coming war on Russia. Much of the evidence previously identified as showing German leaders' last-gasp reluctance to launch full mobilization before caving in to "the pull of military schedules [that] dragged them forward"[93] is now better seen in light of the political and military benefits of having Russia and France make the first moves.[94]

The case of railroads poses a final important anomaly for offense-defense theory. Suppose that offense-defense theorists were correct in claiming that European leaders after 1871 perceived railroads as thoroughly favoring attackers and thus expected future wars to be short and decisive. Why then did no major war occur on the continent for almost fifty years? What, besides the continental balance of power, could explain why an intense and prevalent belief in offensive advantage did not yield the kind of preemptive incentives and destabilizing dynamics witnessed later in 1914? If one accepts the historical characterization of shifting strategic perceptions of railroads frequently advanced by offense-defense theorists and compares it with the actual record of international political behavior, it seems that the standard offense-defense prediction should be reversed: war was more frequent when railroads were believed to favor defense (until 1871) and less frequent when railroads were thought to favor offense (after 1871). In reality, railroads were fairly consistently seen on balance to be defense-enhancing. This view did little to ameliorate security competition and prevent war.

Conclusion

Railroads gave armies an unprecedented degree of mobility. On the strategic level, the emergence of railroads meant that troops could be mobilized and deployed more swiftly than ever imagined, could be moved to fight farther from the homeland than ever before, and could arrive on the battlefield in much better condition, with less equipment, and unburdened by supply columns. On the operational level, railroads greatly hastened the movement of forces between and within theaters of conflict. According to offense-defense theory, significant mobility-improving technologies such as this shift the offense-defense balance toward offense. Offensive shifts in the balance should result in quicker and more decisive victories for the attacker on the battlefield, and the perception of increased offensive advantage should lead states to initiate more conflict.

The historical evidence from the railroad era undermines offense-defense predictions. In terms of military outcomes, the introduction of railroads on balance favored the defender. On the one hand, railroads augmented the attacker's power of initiative by making it easier to assemble and concentrate a decisive force for surprise attack. On the other hand, and more important, railroads greatly improved the defender's ability to quickly respond to attacks, reinforce threatened points, and retreat without suffering a decisive defeat. The short and decisive conflicts in Europe between 1850 and 1871, particularly the wars of German unification, resulted largely from crucial asymmetries in power and skill rather than from the offensive advantages of railroad mobility. Moreover, the evidence from these wars indicates significant problems with employing railroads in offensive campaigns. The American Civil War and World War I clearly demonstrated the strategic defensive benefits of railroads.

In terms of political outcomes, the evidence runs contrary to offense-defense predictions. War occurred more often when railroads were thought to have given defenders a decisive advantage, and less often when railroads were supposedly (mis)perceived to have favored offense. Between 1850 and 1871, although virtually all European military experts (including, most important, those in Prussia who frequently counseled for war) believed that railroads had made it easier to defeat foreign invasion, conflict was plentiful and security scarce. Prussia, which stood to benefit the most from the defensive advantages of railroads, provoked three major wars in a single decade. After 1871, railroads were seen to be vital to any war plan. If, as claimed by some, many Europeans concluded that railroads gave attackers a tremendous boost in war, this was certainly not a view shared by German leaders. When war finally broke out almost half a century later, the statesmen most responsible for generating the conflict were not only fully cognizant of the defensive military advantages of railroads but also hopeful that such advantages offered the key to their offensive political ambitions.

3

The Small Arms and Artillery Revolution

Industrialization transformed military technology and produced a revolution in firepower in the late nineteenth and early twentieth centuries. Significant developments in metallurgy, chemistry, physics, and the system of mass production were applied to the science of war and resulted in exponential growth in the destructive power of small arms and artillery. The development of breech-loading rifles and quick-firing artillery (to replace muzzle-loading, smoothbore guns), in particular, meant that armies were able to deliver far larger quantities of deadly firepower at much greater distances with exceptional accuracy in a fraction of the time previously possible.

Proponents of offense-defense theory make two arguments about the military and political impact of this unprecedented revolution in firepower. First, as expected, proponents contend that the new technological realities gave an enormous advantage to defenders and resulted in longer and more indecisive warfare. Second, proponents claim that despite the objective shift in the offense-defense balance toward defense, European statesmen and military leaders erroneously perceived greater offense-dominance in the succession of new technologies. According to the argument, European leaders believed that attackers would benefit most from vast increases in firepower, thought future wars would be short and decisive, and thus were more inclined to initiate conflict. Moreover, offense-defense theorists contend, these massive misperceptions of technological change—the "cult of the offensive" and "short-war illusion," as they have been labeled—served as a master cause of World War I. "During the decades before the First World War, a 'cult of the offensive' swept through Europe," writes Stephen Van Evera. "Europeans increasingly believed that attackers would hold the advantage on the battlefield and that wars would be short and decisive."[1] Robert Jervis argues that if the participants of World War I had known the defensive advantages of firepower before-hand, "they would have rushed for their own trenches rather than for the enemy's territory."[2] For George Quester, the views of political and military leaders before World War I "simply reflected the predominant expecta-tion on the technological future of warfare, which promised ever greater speed and decisiveness in the execution, ever greater power to the offen-sive."[3]

This chapter offers a crucial test for offense-defense theory. The small arms and artillery revolution and the lead-up to World War I are seen as paradigmatic illustrations of the tragic consequences that can result when technologically determined defense dominance is ignored or misper-ceived by statesmen. Offense-defense theory has previously been criti-cized for relying on World War I as both the principal source for generat-ing hypotheses and the principal empirical test of these hypotheses—a criticism based on the assumption that doing so would normally bias the results of the test in favor of offense-defense theory.[4] Here, however, a vice might be turned into a virtue: contradictory findings in a model case such as this would seriously undermine confidence in the theory as a whole.

In fact, although the empirical evidence yields qualified support for the offense-defense prediction about military outcomes, it offers almost no support for the explanation offered for political outcomes. First, in terms of military outcomes, offense-defense theory is on solid ground in claim-ing that vast increases in firepower wrought by the small arms and artillery revolution favored defenders over attackers. However, confidence in the validity of the firepower-favors-defense hypothesis is tempered by the fact that the defensive advantages lay primarily at the tactical (not operational or strategic) level of warfare and that doctrinal choices (not just techno-logical conditions) significantly shaped the problem of attacking in the face of modern firepower. Second, and most important, offense-defense theory fails to explain political outcomes. To the extent offensive strate-gies proliferated in pre-1914 Europe, the cult of the offensive was not based on perceptions of offensive technological advantages.[5] Military ex-perts, particularly in Germany, understood that technological advances in small arms and artillery had made offense much more costly and difficult.

Yet these accurate perceptions of the military consequences of the firepower revolution did little to dampen German political ambitions.

Military Outcomes

The firepower revolution, as defined here, encompasses the time of tremendous technological change in small arms and artillery beginning in the 1850s, accelerating in the 1880s and 1890s, and culminating in the period after 1900 through World War I. The list of major innovations and improvements that occurred during this dizzying time of technological experimentation includes the advent of the percussion cap and elongated conical bullet; rifled infantry weapons and modern artillery; breech-loading, repeating, and recoil mechanisms; smokeless powder, metallic cartridges, and magazines; machine guns; and high-explosive shells and shrapnel.[6] The combined effect of this firepower revolution was to produce an exponentially more lethal battlefield, where massed frontal assaults by infantry became exceedingly difficult and many conflicts were marked by costly battles of attrition.

Evidence from Major Wars

Few of the most important modern firepower advances were used in the Crimean War (1854–56), but signs of the impact of the emerging technologies were evident even then. Russian forces besieged at the key naval port city of Sevastopol had relatively few riflemen at the time, but these forces were highly effective when fighting on defense. Indeed, for this reason the British and French opted for a protracted siege of the city rather than attempt a quick assault. An exception was the British cavalry charge across open terrain against Russian field guns at the Battle of Balaklava (October 1854), a suicidal assault made famous by Lord Alfred Tennyson's poem "The Charge of the Light Brigade." At the Battle of Inkerman (November 1854), British and French troops armed with new Minié rifles inflicted devastating casualties on assaulting Russian infantry columns attempting to break through the besieging troops. The Russian attackers at Inkerman suffered four times as many casualties as the defenders. In sum, the Crimean War showed that even primitive rifles could be highly effective weapons against attacking infantry.[7]

The defensive advantages wrought by the firepower revolution were readily apparent on the battlefields of the American Civil War (1861–65).

The armies began the war with arsenals consisting primarily of smooth-bore muskets but soon adopted the rifle as the standard infantry firearm. The smoothbore had an effective range of only about one hundred yards, which necessitated the massing of troops into tight formations for attack and defense. The most common rifles, on the other hand, had an effective range of over five hundred yards, meaning that fewer defenders could inflict unprecedented casualties on attacking troops, especially if those troops remained wedded to massed frontal assaults.[8] Unfortunately, both Confederate and Union forces repeatedly sent waves of attacking infantry against entrenched defenders armed with rifles and artillery—and suffered predictable and costly defeats. Among many cases, the Seven Days' battles and the battle at Gettysburg witnessed huge casualties for the attacking Confederate forces; the battles of Antietam, Fredericksburg, Cold Harbor, and Kennesaw Mountain yielded similar outcomes for attacking Union troops. For example, the frontal assault on entrenched Confederate forces at Cold Harbor, Virginia (in June 1864) was called off after Union troops suffered seven thousand casualties in less than an hour.[9]

Prussia benefited greatly from new firepower innovations in the Wars of Unification (1864–71). The previous chapter showed that the quick and decisive Prussian victories against Denmark, Austria, and France should not be attributed to the offensive advantages of railroads but rather to the great asymmetries in military capability and skill between Prussia and her adversaries. In the same way, the striking success of Prussia's offensives in this period should not be taken as proof that firepower somehow favored attackers, as Prussia enjoyed a technical edge in this category as well. In fact, the battlefield evidence from the Wars of Unification supports the conclusion that the emerging firepower revolution favored defenders. By mid-century the Prussian army had adopted the Dreyse "needle gun," a breech-loading rifle, as the basic infantry weapon, while the other European armies continued to employ muzzle-loading rifles. A soldier—lying down—could load and fire the needle gun almost five times as fast as any other infantry weapon at the time. In the war against Denmark, the Prussian needle gun proved devastating on defense against Danish troops advancing in traditional close-order formation and armed only with muzzle-loading rifles. It is worth noting, however, that the Prussians themselves suffered heavily when they launched frontal attacks against even inferiorly armed Danish riflemen, as they did at Dybbøl in 1864.[10] The technical superiority of the Prussian breechloader was even more apparent in the war against Austria in 1866, when the Prussian infantry inflicted enormously lopsided losses on advancing Austrian columns. From the first battles of the war, at Trautenau and Vysokov, to the culminating battle of Königgrätz, the casualty ratio between Austrian and Prussian forces was roughly six to one, close to the fivefold advantage the Prussians had in the rate of fire of their needle gun.[11] By 1870, the French, like all other European armies, had adopted breech-loading rifles and for a brief time at the outset of the war benefited from the defensive advantages of modern firepower. At Gravelotte-St. Privat (August 1870) Prussian forces suffered enormous casualties in close-order frontal attacks against entrenched French defenders armed with highly effective Chassepot rifles.[12] (The Germans suffered eight thousand casualties in three hours.) Ultimately, however, Prussia's loss of technical superiority in infantry armament was more than offset by the triumph of its Krupp steel rifled breech-loading artillery, which quickly devastated the French forces in the field.[13] The quick and decisive outcomes of these wars should not hide the fact that the firepower revolution had greatly strengthened tactical defenders.

The nature of combat in the Russo-Turkish War (1877–78), British-Boer War (1899–1902), Russo-Japanese War (1904–5), and, of course, World War I (1914–18) clearly demonstrated that defenders armed with modern rifles, machine guns, and artillery had gained an enormous advantage against assaulting infantry. For example, the Russians greatly outnumbered the Turkish forces at the siege of Plevna in 1877 but suffered thirty thousand casualties in three failed assaults against the entrenched defenders, who were armed with American breechloaders.[14] In South Africa, a British force that at one point numbered five hundred thousand engaged an adversary that never deployed more than forty thousand men. However, in a war in which both sides employed smokeless repeating rifles, machine guns, and rapid-fire artillery, British assaults against the Boer defenders were repeatedly repulsed with heavy losses, and it took almost three years for the British to claim victory.[15] In Manchuria, in a conflict that witnessed the use of virtually all the technologies of the firepower revolution, the Japanese had seventy thousand casualties in ten days of indecisive fighting in their effort to capture the Russian base at Mukden.[16]

Little needs to be recounted about how defensive firepower dominated the battlefields of World War I. The parity of strength of the opposing coalitions and the combination of modern rifles, machine guns, and field artillery possessed by all resulted in the repeated slaughter of attacking forces and prolonged fighting. By 1916, battles lasted for months on end, often without a clear victor. For example, the series of attacks and counterattacks at Verdun in 1916 lasted for ten months and resulted in roughly seven hundred thousand German and French casualties, with almost half that number killed. On the first day of the battle of the Somme, on July 1, 1916, attacking British troops suffered fifty-eight thousand casualties, including twenty thousand dead. Combined British, French, and German casualties for the entire Somme campaign, which lasted over five months, were well over one million.[17] Costly battles of attrition and stalemate re-curred across the western front in Europe from 1915 through 1917; and most of the total twenty-nine million killed and wounded in the war re-sulted from infantry attacks against entrenched defenders. World War I confirmed that the firepower revolution had increased the supremacy of tactical defense.

Defensive Advantages: Some Qualifications

The nature of warfare in the period under examination supports the offense-defense hypothesis that firepower-enhancing technologies favor defense. However, two significant qualifications should be noted and dis-cussed further. First, the defensive advantages of firepower in this period rested primarily at the tactical, not operational or strategic, level of war-fare. Certainly the difficulty of conducting offensive tactical maneuvers adversely affected larger operational and strategic objectives, especially where geography constrained military operations, but not always. This is an important qualification because ultimately offense-defense theory aims to explain the outbreak of war on the basis of perceptions of the fea-sibility of strategic goals. Second, the choice of doctrine and tactics, not simply the nature of technology, was an important factor shaping battle-field outcomes at the time. In fact, when the prevailing doctrine of massed frontal assaults was modified or rejected, tactical offense became more feasible. At a minimum, this evidence suggests that the defensive advan-tage of firepower in the late nineteenth and early twentieth centuries was not as intrinsic to the pool of prevailing technology as it is often por-trayed.

Tactical Defense versus Operational and Strategic Offense   Consider first the evidence that the tactical impasses created by firepower technologies did not necessarily translate into strategic or operational deadlock. One of the best examples, from the era of the German Wars of Unification, was the method of attack using flanking marches and enveloping operations to place forces where they could employ tactical defensive firepower against enemy forces. This fusion of the strategic offensive with the tacti-cal defensive contributed to Prussia's quick and decisive victories against Austria in 1866 and especially France in 1870. For example, in September 1870, in the decisive battle of the war, German forces maneuvered be-tween Paris and Sedan, thereby enveloping an entire French army at the Franco-German border. The French were compelled to try to break out of the ring but were repeatedly repulsed by German artillery and forced to surrender.[18]

In the American Civil War, Union forces were able to bring the war to an end more quickly after they learned to combine the strategic or operational offensive with the tactical defensive in pursuit of Confederate forces. The Confederates had pioneered the method of attacking by moving infantry into positions where they could employ defensive firepower—for example, at Second Bull Run in August 1862, when forces were marched into the gap between retreating Union forces and Washington—but the Union army used the method to great effect in its partial encirclement of Richmond and Petersburg and pursuit to Appomattox in 1864 and 1865.[19] Similarly, during his offensive campaign on Atlanta in the summer of 1864, General William Tecumseh Sherman repeatedly used entrenched holding tactics to keep the Confederate army on his front while he moved against the enemy flanks.

In the Boer War, the British quickly understood and began to address the offensive problems created by the defensive firepower of small arms. In 1900, they evolved new methods for attack, utilizing firepower to pin Boer forces in place while turning the Boer flanks with horse-mounted infantry.[20] This combination of operational maneuver with tactical firepower resulted in important successful offensives, such as at Paardeberg (February 1900), when a large force of Boers surrendered after being encircled and bombarded by British troops and artillery.[21] Finally, although the Russo-Japanese War made clear the value of firepower on the tactical defensive, the Japanese won the war's most critical battle when they outflanked the Russians at Mukden. In fact, although every major war between 1861 and 1905 saw battles in which defensive firepower halted tactical assaults, all were ultimately decided by strategic offensive maneuvers.[22]

The superiority of defensive firepower reached its apogee on the western front in World War I, but it is at least worth noting the argument that the German strategic offensive through Belgium and France at the outbreak of the war in 1914 almost succeeded. In a matter of weeks, the German armies had conquered huge swaths of French territory, including the most important industrial regions, and were bearing down on Paris. Indeed, the label for France's defeat of the massive German wheeling movement outside Paris was "the miracle on the Marne."[23] Because historians continue to fiercely debate the nature, goals, and even existence of the German war plan, the "Schlieffen Plan," determining whether the German offensive could have succeeded is inordinately difficult.[24] Some argue that the German campaign was fundamentally flawed because of inherent logistical constraints.[25] Others view the plan as strategically and logistically sound, but they fault the younger Helmuth von Moltke because he modified ("marred" or "watered down," are the typically pejorative terms) the plan in the years leading up to 1914. For example, some argue that Moltke's decision to strengthen German forces on the French border instead of the forces destined to wheel through Belgium and Luxembourg prevented French forces from advancing into German territory and, in turn, allowed those forces to retreat to the outskirts of Paris in time to stop the German offensive.[26] Regardless, many agree with the basic conclusion that the German offensive in the west in 1914 was an exceptionally risky and uncertain gamble but also one that brought the Germans close to a quick and decisive victory despite the enormous tactical dilemmas created by the firepower revolution.

The initial war of movement on the western front in World War I settled into a stalemated struggle of trench warfare by mid-September 1914. However, the point that the firepower revolution favored tactical, but not necessarily operational or strategic, defense is reinforced by the nature of the fighting on the eastern front, which remained relatively fluid throughout the war. With a lower force-to-space ratio in eastern Europe, offensive military operations could aim for the kind of turning or envelopment movements that would force the enemy into taking the tactical offensive in order to restore its lines of communication. This is precisely what happened at Tannenberg, in August 1914, where the Germans encircled an entire Russian army with a relatively thin line of forces and capitalized on the primacy of defensive firepower to prevent a Russian retreat.[27]

Defensive Advantages and Tactical Doctrine   The second major qualification to the conclusion that firepower technology favored defense in this period stems from the role that changes in tactical doctrine played in ameliorating the problem of attack. As far back as the American Civil War, attacking infantry learned to reduce casualties by breaking up from waves of attacking troops into smaller groups, which then alternated between advancing and providing covering fire. In the Boer War, the British shifted tactics from massed frontal assaults to dispersed groups rushing from one covered position to the next. In the Russo-Japanese War, the Japanese conducted numerous successful (though still enormously costly) infantry assaults against entrenched defenders using similarly new tactics. The historian Michael Howard describes the tactics: "The Japanese bayonet assaults came . . . at the end of a long and careful advance. They approached whenever possible by night, digging in before dawn, lying up by day, and repeating the process until they could get no further. Then, breaking completely with the European tradition of advancing in extended lines, they dashed forward in small groups of one or two dozen men, each with its own objective, moving rapidly from cover to cover until they were sufficiently close to assault."[28]

The search for a tactical offensive solution was most advanced on the western front of World War I, where the flankless front lines rendered the primacy of tactical defense over frontal attacks unmistakable. Although the search for new offensive tactics began immediately, a workable system of attack—using innovative infantry, artillery, and combined arms tactics—emerged in the later years of the war and led to a series of German and Allied offensive breakthroughs in 1918. The Germany army was the first to master and implement these "infiltration tactics" as a means of breaking the stalemate of trench warfare, but the tactics were eventually incorporated by all sides in the conflict. Such tactics of attack called for a brief surprise artillery bombardment (a "hurricane barrage") aimed at disrupting narrow weak points in the enemy line, followed by small groups of independently moving infantry ("storm troops") seeking to bypass points of strong resistance and advance as far as possible. Ironically, firepower was a crucial element of the new attack doctrine, as the infiltrating troops were thoroughly equipped with light mortars and machine guns, hand grenades, and flamethrowers and supported by other units providing constant suppressing fire during the advance.[29]

The Germans began to employ infiltration tactics in late 1917 and used them with great success in the Ludendorff offensives in the spring of 1918.[30] Despite no major changes in technology or weaponry, each of the three Ludendorff offensives achieved significant and unprecedented breakthroughs followed by deep advances behind the Allied lines. Stephen Biddle, in an in-depth case study of the first of these offensives, convincingly shows that the new infiltration tactics, and not any numerical imbalance of power or new weapons technology, were the source of Germany's ability to break the tactical stalemate.[31] These offensives ultimately failed on the strategic level because the Germans lacked the transportation and logistical capabilities necessary to follow up on their tactical successes. The Allies, on the other hand, put similar tactics to good use in their own offensives through the end of the war.[32] Clearly, massed frontal assaults in the face of modern firepower were not the optimal method of attack; more difficult to conclude, however, is how World War I would have looked if infiltration tactics had been employed at the outset. At a minimum, the series of offensive breakthroughs in 1918 is powerful evidence that the defensive stalemate on the western front was not as intrinsic to prevailing firepower technology as is often portrayed.

In sum, the idea that the firepower revolution ushered in an era of long and indecisive warfare has validity but requires important qualification.

Modern firepower enhanced the power of defense at the tactical level, but this did not always translate into operational and strategic stalemate. In addition, tactical innovation—not technological change—ultimately broke the stalemate created by defensive firepower. These qualifications notwithstanding, the battlefield evidence from this period supports the hypothesis that firepower-enhancing technologies shift the offense-defense balance in favor of defenders.

Political Outcomes

Offense-defense proponents argue that Europeans misperceived the objective shift in the offense-defense balance wrought by the firepower revolution, erroneously believed that prevailing technologies favored the attacker, concluded that future war would be quick and decisive, and thus adopted policies and strategies that made the outbreak of war more likely. Van Evera writes, "During the decades before the First World War, a 'cult of the offensive' swept through Europe. Belief in the power of the offense increased sharply after 1890 and rose to very high levels as 1914 approached." He contends that Europeans "largely overlooked the lessons of the American Civil War, the Russo-Turkish War of 1877–78, the Boer War, and the Russo-Japanese War, which had revealed the power of the new defensive technologies."[33] Thus, Quester notes, World War I was "launched on the illusion of offensive advantage."[34] Much of the "short-war illusion" thesis is based on the view that Germany went to war in August 1914 armed with a bold operational plan for quickly enveloping and annihilating the French army, shifting the victorious troops to the east against Russia, and securing victory in Europe by Christmas. Summing up the conventional view, one author writes that "forecasts of a short war were a monumental strategic miscalculation that had the most tragic consequences."[35]

The historical record is at odds with these claims. Records and documents recovered in the last decade from the former East German military archive form the basis of a new historical interpretation that undermines the conventional wisdom of a "short-war illusion." The new interpretation—called the "Förster thesis," after the German historian Stig Förster, who was one of the first to explore the recovered archival materials—argues that Germany's senior military leaders were remarkably uniform in their view that the next European war would likely be a protracted struggle. Reinforcing the findings of Terence Zuber and other historians who have investigated the new documentary materials, the Förster school also contends that the Schlieffen Plan was little more than a proposal for how to win the initial frontier battles of an expected long, hard war. The view that war would be long and indecisive was based in large part on the knowledge that the firepower revolution had strengthened the defender and made tactical stalemate more likely.[36]

The remainder of this chapter discusses the evolution of German perceptions of the nature of the firepower revolution and how these perceptions interacted with strategic planning and decision making to either promote or dampen conflict. The focus on German perceptions (rather than those of other Europeans) is justifiable because Germany was principally responsible for provoking World War I, as well as the earlier Austro-Prussian and Franco-Prussian wars, the only other major wars on the continent in the previous half century.[37] (Thus, offense-defense claims should find strong support in the record of German decision making addressed here.) From before the Wars of Unification through World War I, most German military officers and political leaders accurately perceived that new firepower innovations conferred a marked superiority on tactical defenders. After 1864, in particular, German leaders expected warfare to be far costlier for attacking forces because of the tactical problems created by lethal firepower. However, this view did not dissuade Germany from attacking or provoking war with its neighbors in 1866 and 1870 or in 1914, by which time the superiority of defensive firepower had become manifest to all. To the contrary, given Germany's strategic dilemma and desire to secure its position of dominance in Europe, it was precisely the awareness that war could easily become long and indecisive that compelled Germany to seek a plan for rapid victory. That Germany went to war in 1914 without having secured that plan is all the more reason to question the core claims of offense-defense theory.

Perceptions of Firepower and the Wars of Unification

Long before 1914, the Prussian general staff strove to understand the strategic and tactical implications of the plethora of technological advances of the era. Prussian officers were more actively involved in the study of military history and contemporary campaigns and battles than those of any other army at the time. Indeed, the general staff, under the leadership of Helmuth von Moltke (the elder), was a model of professional military expertise that would eventually be copied by armies worldwide.[38] Not surprisingly, Prussian military leaders were quite cognizant of the defensive potency of the new firepower technologies by the 1860s, if not before. Historian Dennis Showalter notes, "Beyond doubt, the [Crimean War] demonstrated the growing importance of firepower on a modern battlefield. It was generally accepted after 1854 that the introduction of long-range rifles would strengthen the defense."[39]

Moltke's understanding of the tactical advantages of defensive fire-power was crystallized by what he learned about the American Civil War and, especially, his experience in Prussia's invasion of Denmark in 1864. In the latter conflict, Moltke witnessed firsthand both the effects of Prussia's breech-loading needle gun against attacking Danish troops and the costly attacks by Prussian forces at the battle of Dybböl. Moltke concluded that in the age of the rifled firearms, no combination of bravery and superior numbers could overcome the problem of attacking frontally against well-armed defenders. The evidence that he held this view is overwhelming: in 1863, he sent an observer to the American Civil War, who witnessed fourteen battles in six months of that year (including the fighting at Chancellorsville and Gettysburg) and reported back to Moltke and the general staff about the tremendously lethal impact of artillery and rifled small arms on attacking infantry.[40] In 1864, in a special issue of the official journal of the German general staff, Moltke published an essay, "Remarks on the Influence of Improved Rifles on the Attack," in which he argued that direct, close-order frontal attacks were a mistake and should be avoided at all costs.[41] In an 1865 essay, Moltke wrote: "Open terrain . . . assures the defense of a strong position because of the effects of modern firearms. A direct advance has little hope of success against such a strong position." In the same essay, he noted, "All the above shows that a direct advance across an open front cannot succeed, and that the attack, if possible, must be directed against the defender's flanks and must seek out covered and broken terrain to allow the attacker to bring his fire into effect and enable him to advance gradually."[42]

In 1869, Moltke produced a new set of guidelines, "Instructions for Large Unit Commanders of 24 June 1869," which provided the basis for German operations for the next seventy years. In light of the popular belief that Europeans were ignorant of the objective offense-defense consequences of new firepower innovations, Moltke's training manual is worth citing at length:

> In the next war our needlegun will not again be opposed by a far inferior rifle, but on the contrary, an entirely equal weapon. . . . [Under such conditions] infantry has the power to repel any frontal attack with its rapid fire, even that of the most audacious opponent.
>
> This conviction must be awakened and ingrained in the infantry. The infantry may assure itself that it cannot be attacked from the front and only has to be slightly concerned if it turns its back. An infantry [unit] whose flank is covered, which does not have to pay attention to losses

from long-range fire, and which delivers a cold-blooded volley fire against the enemy's sometimes traditional dare-devil attacks, is invincible.

It cannot be denied that a man firing while stationary has the advantage over the advancing firer . . . and that if calm steadfastness opposes zealous élan, the effects of fire that have become so much greater in our time decide the issue.[43]

Of course, the German general staff did not passively accept this condition of tactical defensive superiority. As Showalter writes,

Reduced to its basic terms, the challenge they faced was this: an increasing number of Europe's infantrymen were carrying rifles which could kill at five times the range of the smoothbore musket. Yet no one had devised a way to make men five times braver, five times as willing to die, or even five times quicker on their feet. . . . [T]he essential question in those years was not *if* the infantry could attack in the face of modern firearms, but *how* it could do so without crippling losses.[44]

One potential solution, offered by Moltke as early as 1858, was that Prussian troops should undertake flanking maneuvers to force any potential enemy into taking the tactical offensive, employ defensive firepower to devastate the attacking forces, and then launch a counterattack to defeat the remaining enemy forces. "Our strategy must be offensive, our tactics defensive," Moltke famously proclaimed. As discussed earlier in this chapter, the new doctrine of combining strategic offensive mobility with tactical defensive firepower was used with great success against Austria in 1866 and France in 1870.

German political leaders were not oblivious to the military conditions created by enhanced firepower technologies. Moltke's rigorous military analysis was known and respected by Chancellor Otto von Bismarck, the two had a good professional relationship at the time, and both leaders coordinated strategic and diplomatic objectives. In short, the view of the increased power of the defense did not dissuade Prussia from initiating war in 1864, 1866, and 1870. Moltke clearly recognized the changes wrought by modern firepower and understood that these changes had strengthened the defensive, but he nonetheless adapted military doctrine to suit larger offensive operations and objectives. To be sure, Prussian aggression was driven by Bismarck's desire to expand Prussia's borders and create a unified Germany. However, the belief that technology rendered offense exceedingly difficult did little to ameliorate these ambitions.

Perceptions of Firepower between the Wars of Unification and World
War I

It is no longer controversial to argue that German leaders were principally responsible for starting World War I. The crux of the issue here is
whether misperceptions of the offensive advantage of military technology
dominated German thinking in the years leading up to the war and
whether these views made war more likely. In fact, German military leaders were fully aware of the defensive superiority conferred by the firepower revolution. The German general staff evaluated the battlefield consequences of the small arms and artillery revolution more objectively than
did all other European general staffs at the time. The military and civilian
leadership understood that warfare would be costly and difficult; even the
cleverest offensive plan could easily devolve into protracted battles of attrition. These perceptions did not dampen Germany's offensive ambitions, which aimed at breaking its encirclement by France and Russia
(and, eventually, Britain), fighting a preventive war to forestall the rise of
Russia, and ensuring German hegemony in Europe for the foreseeable future. Instead, German planners shaped their military strategy to accord
with political aspirations. In doing so, they sought to overcome the tactical
problems of firepower through a strategy aimed at achieving a quick and
decisive victory. In short, Germany considered offensive operations extremely difficult, but also absolutely necessary for national policy.

The period between 1871 and the early 1900s was one of constant feud-
ing within the various branches of the German army over the tactical implications of modern military technology. Conservatives and traditionalists were aware of the devastating effects firepower had on attackers—for
example, at Gravelotte-St. Privat in 1870—but rejected the loss of battlefield command and control that would occur if the older massed infantry
shock tactics were replaced by newer dispersed formations with infantry
seeking cover and advancing more cautiously. For traditionalists, the tactical solution to the challenge posed by the firepower revolution lay in the
moral strength and spirit of Germany's soldiers. Pragmatists and reformers, such as Moltke, understood that the bloodlettings of the previous
decade could be reduced only through tactical reforms and new infantry
regulations that called for greater troop dispersal, mobility, and flexibility. Although the traditionalist views prevailed by the late 1870s, the tactical reformers gradually ascended afterwards as the German army absorbed the lessons of the Russo-Turkish War, British-Boer War, and
Russo-Japanese War.[45]

Although the proper tactical response to the firepower revolution was
hotly contested within the army in the decades leading up to World War I,

German views of the broader operational and strategic solutions were surprisingly consistent from the time of Moltke's reign as chief of the general staff through that of his successors Alfred von Waldersee from 1888 to 1891, Alfred von Schlieffen until 1905, and Moltke the younger up to 1914. The central predicament for German strategic planning was that the country was geographically situated between an implacably hostile France to the west and her likely ally Russia in the east, presenting Germany with the prospect of a two-front war against a coalition with greater combined resources and total manpower. The only way to win such a war would be to quickly defeat one adversary before turning on the other. Prolonged fighting on either front would prove too much for Germany to handle and had to be avoided at all costs.

Unfortunately for Germany, Moltke and most of his fellow officers realized after 1871 that quick and decisive victories on a par with those won in the Austro-Prussian and Franco-Prussian wars would be increasingly unlikely in the future. In the war with France, both French cavalry and Prussian infantry assaults against defenders had suffered horribly. In an 1871 study that considered the possibility of a war on two fronts, Moltke wrote, "Germany cannot hope to rid herself of one enemy by a quick victory in the west in order then to turn against the other. We have just seen how difficult it is to bring even the victorious war against France to an end."[46] Moltke elaborated on his pessimism in an 1874 essay, in which he wrote, "I am convinced that improvements in firearms have given the tactical defense a great advantage over the tactical offense. It is true that we were always on the offensive in the campaign of 1870 and that we took the enemy's strongest positions. But with what sacrifice! Taking the offensive only after having defeated several enemy attacks appears to me to be more advantageous."[47] Moltke recognized that offensive warfare would become even more difficult as Germany's rivals became stronger and better armed with the new weapons of firepower. Even operations that sought to fuse the strategic offensive and tactical defensive, Moltke concluded, would likely deliver only limited victories and military stalemate. For these reasons, Moltke's plans concentrated on a limited offensive in the east to take Poland away from Russia, so as to form an eastern strategic buffer to compliment Germany's previous seizure of Alsace-Lorraine as a western buffer. If forced to fight a two-front war, Germany would ride out the French offensive and then look for the best opportunity to counterattack. With some variation from year to year, this vision remained the basis for German operational planning from the early 1870s until 1887. In that year, under pressure from Bismarck, whose first priority in the event of war was to eliminate the French threat to Germany's position of power, and with the encouragement of Moltke's soon-to-be successor, Waldersee, who feared the growing strength of the French forces, Moltke adopted a plan to commit the mass of the German army in the west against France. Despite Moltke's definitive switch from a "Russia first" strategy to a "France first" strategy—which has previously been incorrectly identified with Schlieffen—neither Moltke nor Waldersee was able to develop a workable offensive plan for achieving quick and decisive victory against France. The only apparent options for Germany were to launch a frontal attack across the Franco-German border or to rely on a classic defensive-offensive operation to stymie a French attack, neither of which was expected to yield more than frontier victories. The more likely scenario, given the growing prospect of Russian intervention in a Franco-German war, would be a long and arduous war of attrition.[48] Thus Moltke's warning to the German Reichstag in 1890 about a future war lasting seven or even thirty years: "Woe betide him who sets Europe ablaze."[49]

The idea that Schlieffen, chief of the general staff from 1891 to 1905, drew up the perfect operational solution for Germany's two-front-war dilemma is not credible. Although the subject continues to fuel much historical scholarship and debate, the full origins and nature of the Schlieffen Plan need not be addressed here.[50] The point worth making is that Schlieffen and his staff shared his predecessors' realistic appreciation of the nature of modern war and recognition of the high probability of a protracted conflict in Europe, even as he carried on the search for an operational means for rapid and decisive victory. The highest priority remained to seek some kind of victory in a swift offensive in the west against France before turning to deal with Russia. As France and Russia solidified their military alliance between 1891 and 1894 and built up their border fortifications, the prospect of a two-front war became both more likely and less attractive. The conundrum was not just that Germany would be confronted by numerically superior adversaries that could mobilize faster than ever before but also that the firepower revolution was undermining Germany's ability to conduct quick offensive operations. An 1895 general staff study concluded that a frontal attack across the Franco-German border, as planned by both Moltke the elder and Schlieffen, would be slow and costly. "We cannot expect quick, decisive victories," advised Schlieffen's quartermaster-general, Ernst Köpke. "Even with the most offensive spirit . . . nothing more can be achieved than a tedious and bloody crawling forward step by step here and there by way of an ordinary attack in siege style."[51]

Schlieffen's recognition of the defensive advantages of modern firepower and fortifications provided the entire rationale for considering an alternative offensive that would aim at the French army's flank and rear by moving through Belgium and Luxembourg. From 1897 to 1905, Schlieffen and his staff explored variations on the plan through annual staff rides, war games, and continuous analysis. Rarely were the results satisfactory: often the French side won an outright victory or the German advance got bogged down in frontal battles. Lessons from contemporary conflicts, such as the Russo-Japanese War, that demonstrated the effects of defensive firepower were also troublesome. As Schlieffen wrote in 1905, "If the enemy stands his ground . . . all along the line the corps will try, as in siege-warfare, to come to grips with the enemy from position to position, day and night, advancing, digging in, advancing. . . . The attack must never be allowed to come to a standstill as happened in the war in the Far East."[52] Moreover, Schlieffen warned, "if the French . . . retreat behind the Marne . . . the war will be endless."[53] Leaving aside, again, the debate over whether the Schlieffen Plan left to Schlieffen's successor, Moltke the younger, in 1905 constituted a design for a rapid and massive envelopment movement through the Low Countries, around Paris, and culminating in a giant battle of annihilation or, more persuasively, was simply the latest version of a plan dating back to Moltke the elder for how Germany might win the initial frontier battles of a longer war, it is clear from the evolution of Schlieffen's studies and analysis that he never devised a workable strategy for winning a quick and decisive war against France, much less a solution to Germany's two-front-war dilemma. The consequences of that failure, however, were unacceptable: Germany could not win a prolonged struggle on several fronts against the superior numbers and resources of its adversaries. Thus, even after retirement, Schlieffen continued to advocate a rapid campaign of annihilation in the west because he refused to accept the implications of a long war of attrition for Germany's future.[54]

Recently recovered documents show that the younger Helmuth von Moltke, chief of the general staff from 1906 through the outbreak of World War I, fully understood that the Schlieffen Plan faced a number of serious operational problems and had major doubts that any future war would be short.[55] Unfortunately, as Hew Strachan notes, "[h]e combined two qualities—fatalism and intelligence—which not only stood uneasily with each other, but also were ill-adapted to the exercise of decisive leadership."[56] In 1905, upon being offered his top position, Moltke told Kaiser Wilhelm II that even a one-front war against France would be grim: "We will not anymore, as in earlier times, be faced with a hostile army that we can engage with superiority but with a nation in arms. It will be a people's war that cannot be won in one decisive battle but will turn into a long and tedious struggle with a country that will not give up before the strength of its entire people has been broken. Our own people too will be utterly exhausted, even if we should be victorious."[57] In letters to his wife he re-

ferred to "a murderous European war" and a "general European mas-
sacre, at whose horror one could only shudder."[58]

Moltke's words of caution were not unique; military writers and planners before 1914 assumed that if a quick and decisive victory were not possible, war would ultimately become apocalyptic and destroy masses of
people, resources, civil society, and ultimately leaders as well.[59] As Schlieffen feared, "A strategy of attrition cannot be employed when the support
of millions requires the expenditure of billions."[60] Yet Moltke, prominent
officers in and outside his staff, and other governmental officials increasingly believed that a long war on two fronts was the most likely scenario
and began to prepare for a protracted campaign. In 1910, the intelligence
section of the general staff warned that the French armies could not be
defeated quickly or completely. In 1912, Colonel Erich Ludendorff (then
head of the mobilization section of the general staff) and Moltke wrote to
the war ministry warning it to prepare for a lengthy conflict: "We will have
to be ready to fight a lengthy campaign with numerous hard, lengthy battles until we can defeat [even] *one* of our enemies. . . . The need for a
great deal of ammunition over a long period of time is absolutely critical."[61] Such dire warnings, both private and public, were expressed by virtually all of the top military leaders through the outbreak of war. As Holger Herwig writes, "The 'men of 1914' were united in the belief that a
general European war would be anything but short."[62]

The view of Moltke and most military thinkers that war would almost
certainly be long, costly, and grim did not lead Moltke to abandon Germany's offensive plans or to inform the political leadership of his pessimism. Like his predecessors, Moltke accepted the unavoidable fact that
modern warfare would be enormously costly, but he was unable to come
to terms with the implication that war was no longer a feasible option for
German foreign policy. Instead, Moltke and his colleagues sought ways to
maximize Germany's chances of winning whatever kind of war material-
ized, even if it would likely be a long and protracted conflict. Ironically,
Moltke's realistic expectation of a long war of attrition led him to make
changes to the German war plan that arguably reduced the chances of any
kind of quick and decisive victory. Specifically, by shifting the planned advance of German forces on the right wing so as to avoid Dutch territory,
Moltke hoped to preserve German access to outside markets during the
war. "It will be very important to have in Holland a country whose neu-
trality allows us to have imports and supplies. She must be the windpipe
that enables us to breathe," Moltke wrote in 1911.[63] This change, however,
created a serious logistical problem in forcing two armies to advance
through a tight bottleneck at Liège in Belgium, just south of the Dutch
border.[64] Whatever operational changes were made, Moltke, on the eve of

war, was clearly not blinded by a short-war illusion but, astonishingly, was one of the most vocal advocates of launching war as quickly as possible. On July 29, 1914, Moltke sent a letter to Chancellor Bethmann Hollweg in which he discussed the coming "world war," warned of "the mutual tearing to pieces by Europe's civilized nations," and predicted that the war would "destroy civilization in almost all of Europe for decades to come." Nevertheless, Moltke demanded that the chancellor start the war as soon as possible, a plea voiced to the kaiser the same day. Bethmann Hollweg, too, equated the coming war with "the overthrow of everything that exits."[65]

We now know that German military leaders eagerly advocated war in 1914 fully expecting a long and protracted struggle. They had nothing more than a glimmer of hope for a quick and decisive victory. Realistically, the best chance in launching their attack was to win several important battles on the frontiers, which would give Germany some additional strategic advantages before the expected war of attrition. Moreover, German military and civilian leaders agreed on almost all elements of prewar decision making and policy.[66] Both groups believed that a continental war was inevitable, necessary for preserving and extending Germany's power position in and beyond Europe, and better fought now than later—all despite the shared vision of an apocalyptic war. In short, Germany made a colossally tragic gamble in 1914, but it was not a gamble "launched on the illusion of offensive advantage." Many soldiers and statesmen, throughout Europe, anticipated that the next European war would be long and extremely bloody. This view was based in large part on the widespread appreciation of the implications of the firepower revolution.[67] German leaders continuously sought a military formula for a quick and decisive victory but were willing to provoke a war to pursue their political objectives in the absence of one.

Conclusion

The second half of the nineteenth century and early decades of the twentieth century witnessed an unprecedented revolution in military firepower, based primarily on rapid advances first in small arms and then in artillery. The battlefield became a much deadlier place, as armies were able to deliver far larger quantities of firepower at much greater distances with exceptional accuracy. According to offense-defense theory, great improvements in firepower shift the offense-defense balance toward defense. Defensive shifts in the balance, in turn, result in longer and more indecisive warfare. Moreover, perceptions of defensive advantage allow states to feel more secure and pursue less bellicose policies.

The evidence from the period of the small arms and artillery revolution offers at best mixed support for offense-defense predictions. On the battlefield, advances in firepower made tactical offense more costly and difficult. Infantry assaults against defenders armed with increasingly sophisticated rifles, artillery, and machine guns were often repulsed with heavy casualties to the attacker. The nature of combat from the Crimean War through World War I shows the defensive advantages of enhanced firepower technologies. However, this finding about military outcomes needs to be qualified for two reasons. First, defensive tactical advantages did not necessarily translate into operational deadlock or strategic stalemate. The best illustration of this is the classic strategy of combining operational maneuver on offense with tactical firepower on defense, which Germany employed in achieving quick and decisive victories in the Wars of Unification. Second, the tactical impasse created by enhanced firepower was broken through doctrinal innovation, not new developments in technology. These points indicate that the defensive advantage of firepower rested primarily at the tactical level of warfare and was not as intrinsic to the prevailing technology as is often portrayed.

In terms of political outcomes, offense-defense theory fares poorly. Prussian leaders were well aware of the increasing difficulty of attacking in the face of modern firepower. Nevertheless, they initiated or provoked three wars in less than a decade (from 1864 to 1871). German leaders did not allow their spectacular offensive successes in the Wars of Unification to blind them to the defensive impact of firepower. Military observers, including those from Germany, who studied conflicts from the American Civil War through the Russo-Japanese War, understood the growing lethality of firepower and the challenge this posed for carrying out offensive operations. "Nobody was under any illusion, even in 1900, that frontal attack would be anything but very difficult and that success could be purchased with anything short of very heavy casualties."[68] Indeed, actual German war planning from 1871 to 1914 reflected such concerns and was marked by persistent efforts to find an operational solution that would make decisive attacks possible. However, German soldiers and statesmen were not deterred by their failure to find such a solution and recognition that a coming European conflict would likely be a protracted war of attrition. With full knowledge of the huge risks and almost certainly appalling costs of war, they seized what they thought was their last chance to preserve and extend German power in Europe. The belief that modern firepower greatly strengthened defense did little to derail Germany's aggressive foreign policy ambitions or prevent the outbreak of World War I.

4

The Armored Revolution

The character of land warfare was transformed between World War I and World War II by the mechanization and motorization of armies. The most important technological innovation in this period was the tank. The tank's combination of technological advances in the internal combustion engine, armored protection, and radio communication greatly increased operational mobility on the battlefield.

Offense-defense theory contends that large improvements in mobility render offensive military operations more effective. When the offense-defense balance shifts toward offense, attackers are more likely to win quick and decisive victories. According to the theory, the prospect of rapid offensive success increases the probability of war because potential adversaries will be more inclined to initiate conflict in pursuit of security than to risk fighting on the defensive.

Proponents of offense-defense theory believe that the incorporation of tanks into the European armed forces in the interwar period resulted in greater offense dominance. According to Stephen Van Evera, "During 1919–45 the power of the offense was restored by motorized armor and an offensive doctrine—blitzkrieg—for its employment."[1] Robert Jervis concurs with Quincy Wright's assessment: "The German invasion in World War II . . . indicated the offensive superiority of highly mechanized armies in the field."[2] Sean Lynn-Jones writes, "The tank, for example, is useful for offensive and defensive purposes, but without tanks, blitzkrieg offensives

would be virtually impossible. Tanks make it possible for states to launch offensives using large armored formations. In other words, they make offensive strategies far less costly than they would have been without tanks."[3] Charles Glaser and Chaim Kaufmann agree that "tanks . . . are generally considered to have favored the offense."[4]

This chapter evaluates offense-defense theory by exploring the military and political consequences of the armored revolution. Two questions are relevant for examining the case of tank technology. The first question concerns *military outcomes:* What impact did tanks have on the offense-defense balance? Specifically, did tanks in this period strengthen offensive operations relative to defensive operations? The evidence does not support this proposition. In World War I and the interwar period tanks had a negligible effect on operational outcomes. For the evidence of armored warfare in World War II, this chapter considers two separate phases: 1939 through 1941 and 1942–43 to 1944. Military operations in the first period yielded quick and decisive victories for the attacker. However, the evidence shows that this advantage resulted from German material and doctrinal superiority rather than the balance of military technology. The second period provides a better test of offense-defense predictions about tanks because all sides in the conflict were by then relatively adept at armored warfare. The record of military operations in this period does not demonstrate the offensive-superiority of tank forces.

The second question addresses *political outcomes:* Did a belief in the offensive superiority of the tank make war more likely? Specifically, to what degree did perceptions of offense dominance embodied in armored technology encourage Germany to attack her neighbors and launch World War II? Van Evera argues that Adolf Hitler felt free to pursue his vast expansionist aims in large part because of his recognition (unshared outside Germany) that armored technology, combined with an appropriate doctrine for its use, gave offense a great advantage over defense.[5] Specifically, Van Evera claims that Hitler refrained from invading France during the fall and winter of 1939–40 because he had not fully come to grips with this offensive shift and decided to attack France in May 1940 only after he had grasped the offensive promise of the armored blitzkrieg doctrine.[6] However, the evidence shows that perceptions of the offensive advantage of armored warfare were not nearly as important a factor in the outbreak of World War II as proponents of offensive-defense theory suggest. During the interwar period, few military observers in Europe—and more in the Soviet Union than in Germany—viewed the tank as a revolutionary offensive weapon. Even more important, Hitler, contrary to popular lore, was bent on military conquest well before any radical revelation of the offensive power of tanks. The claim may seem controversial, but the evidence

shows that Hitler both attacked Poland in 1939 and decided to attack France in late 1939 and early 1940 well before the emergence of a bold new plan based on new armored tactics and strategic surprise.

Military Outcomes

The impact of tanks on the offense-defense balance is best discussed chronologically: in World War I and the interwar period, in World War II from 1939 through 1941, and in World War II from the winter of 1942–43 through 1944. In World War I and the interwar period, tanks had no discernible effect on the offense-defense balance. In World War II, the most relevant evidence also shows that tanks did not ultimately shift the balance toward offense.

Armored Warfare in World War I and the Interwar Period

The use of tanks in World War I provides little guidance as to their eventual impact on warfare or their effect on the offense-defense balance. All of the major European armies had experimented with armored fighting vehicles by 1914, but only the British and French sought to produce large numbers of tanks by 1916. Tanks were first used in the battle of the Somme in September 1916, achieved their greatest success at the battle of Cambrai in November 1917, and played a large role in the Amiens offensive in August 1918. Tanks occasionally contributed to tactical breakthroughs and penetrations, especially when they were employed together in large numbers over favorable terrain and supported by advancing infantry (as at Cambrai). However, tanks ultimately could not translate any tactical successes into operational victories. The tanks of World War I were usually manned by untrained crews, easily separated from the advancing infantry, often forced to disperse, vulnerable to defensive artillery fire, and prone to mechanical breakdown. In fact, the most spectacular breakthroughs of the war, such as the 1918 Ludendorff offensives, were made possible more through the introduction of new infantry tactics than by tanks.[7] In short, tanks did not have a decisive impact on military operations in World War I.

The twenty years between the two world wars produced little actual battlefield evidence indicating the offensive tactical, operational, or strategic significance of tanks. The wars fought between 1919 and 1939 were primarily either civil wars or colonial conflicts and involved unevenly matched adversaries or forces that were not well equipped with tanks.

When tanks were used in battle, as in the continuing French effort to extend control over Morocco (1908–34), the Italian war against Ethiopia (1935–36), the Spanish Civil War (1936–39), and the Russo-Japanese border clashes in Manchuria (1938–39), the results were not illuminating.[8] In Morocco, for example, French tanks were of limited value in the key battles of 1925 and 1926 because the French had already occupied much of the terrain where tanks might have been decisive. Italian tanks played a greater role in Ethiopia, but the Italian conquest was virtually a foregone conclusion given the lopsided balance of military capability between the sides. In the Spanish Civil War, both Germany and the Soviet Union supplied tanks to the Nationalist and Loyalist forces, respectively, but tanks were indecisive in battle and had little effect on the course of the war. Various armored offensives were repeatedly halted by mountainous terrain, defending infantry, or air power or were unsuccessful because of insufficient numbers of tanks, the failure to deploy tanks en masse, or the subordination of tanks to infantry.[9]

The Soviet Union and Japan engaged substantial tank forces in two major battles—at Lake Khasan (Changkufeng) and Khalkhin Gol (Halha River)—on the border of Manchuria and Mongolia in 1938 and 1939. Evidence for assessing the offense-defense impact of armored warfare is mixed. On the one hand, in both battles, attempted tank-led assaults were repeatedly halted. In July and August 1939, the Japanese army launched two major offensives against the Red Army at Khalkin Gol, near the village of Nomonhan. Although initially successful, these attacks stalled because of superior Soviet armored forces (the Japanese were attacking with dozens of tanks, whereas the Russians had hundreds for defense and counterattack) and antitank weapons. Soviet counterattacks with tanks and other armored cars reversed Japanese gains but resulted in the destruction of many Russian tanks by Molotov cocktails, antitank guns, and antitank mines. The Japanese (and to a lesser extent Soviet) armored offensives also suffered from ineffective tactics, with tanks operating essentially alone and without the support of infantry and artillery to counter enemy antitank defenses. On the other hand, the Red Army forces, led by General Georgi Zhukov, eventually defeated the Japanese at Khalkin Gol with a tank-heavy combined-arms attack that penetrated the defensive line at the flanks, pushed through to the rear, and carried out a double envelopment of a substantial portion of the Japanese forces. This apparent harbinger of offensive tank warfare, however, should be qualified—as it was at the time by participants and afterward by military staffs—in light of the gross imbalance of forces (in tanks, manpower, and antitank guns) and Japan's depleted and exhausted forces (fighting far from home territory with long lines of communication).[10] In sum, the military conflicts in the interwar period do not give much indication of the significance of tank warfare and provide little support for the argument that tanks had decisively shifted the offense-defense balance toward offense.

Armored Warfare in World War II, 1939–41

Germany's rapid envelopment of Polish forces in 1939, quick and decisive defeat of France in 1940, and invasion of the Soviet Union in 1941 were the first significant demonstrations of the battlefield impact of tanks. Germany's devastating attacks with armored and motorized forces at the outset of World War II suggest that the balance of military technology had become offense-dominant. The standard argument is that the attacker had gained an enormous advantage because the inherent mobility of the tank (along with its motorized and mechanized logistical and infantry support) allowed rapidly concentrated forces to break through defender weak points, penetrate deep into the enemy rear, and either encircle or paralyze enemy forces.

The campaigns of 1939–41 offer inadequate evidence that tanks conferred a decisive advantage on the offense, however, because the victories resulted from German material and doctrinal superiority rather than from the balance of military technology. The largest obstacle to assessing the offense-defense impact of any particular weapon usually lies in the need to evaluate outcomes on the battlefield while controlling for different levels of military skill and resources between adversaries. This problem looms large in the case of German offensives at the outbreak of World War II.

First consider the Polish campaign. In September 1939, German forces were better trained, better equipped, and far larger than the Polish army. The Germans sent approximately 1.5 million soldiers to fight against a Polish army that had only about one-third of its fully mobilized force of 1 million available when the invasion began. The Germans also held significant advantages in quality of equipment, communications, and logistics and in numbers of heavy artillery and other fire support weapons.[11] Any evaluation of the use of armor in the attack would be limited by the fact that Germany had a far greater number of tanks (over thirty-one hundred against about three hundred Polish tanks), the Polish had few antitank guns, and the German infantry and air forces made relatively more important contributions to the war's outcome. The Luftwaffe (with 2,085 planes), for example, destroyed the Polish air force (of 313 aircraft) in the first week of the war and thereafter bombed Polish forces at will. Moreover, the official German report on the Polish campaign gave the infantry the main credit for the victory.[12] In short, although German armored and motorized divisions played an important role in overwhelming the Polish front lines and encircling Polish forces, the German victory was a foregone conclusion only hastened by Polish weaknesses and mistakes.[13]

Germany ran roughshod over France in May 1940, attaining one of the quickest and most decisive victories in military history. In this case, the German armored forces were neither more numerous nor technically superior to Allied forces. In fact, deployed forces on the western front on May 10, 1940, show a slight numerical superiority for the Allies in tanks (at a ratio of 1.3 to 1) and manpower (1.2 to 1).[14] In addition, the French possessed better medium and heavy tanks and a huge advantage in artillery (11,200 guns to 7,710) and munitions.[15] Germany's swift victory in the face of these disadvantages has contributed to the notion that armored technology provided the key to offensive success, but that is mistaken.

The root cause of the German victory over France lies with Germany's superior strategy, tactics, and organization, rather than in the nature of its military hardware. The details of Germany's superior doctrine and the Allies' glaring weaknesses are well known. The German army strove for independent, flexible, and mobile fighting units that could exploit opportunities as they arose on the battlefield, while the Allies (especially the French) preferred their units to fight in a controlled, methodical fashion and stressed the importance of fixed defenses and fircpower over defense in depth and maneuver. The Germans concentrated their armored forces into powerful panzer divisions that were followed by fully motorized infantry divisions, while the Allies dispersed their tanks among the slower infantry and artillery forces. Germany achieved complete strategic surprise by attacking through the supposedly impenetrable Ardennes forest with an armored force five times as large as that of the French defenders, while the French ignored many intelligence indicators of the German concentration and were woefully under-equipped with antitank defenses opposite the Ardennes. Finally, the Allies made a disastrous advance eastward into Belgium with their best mobile divisions to occupy defensive positions, while the German panzer divisions were concentrating in the main area of attack to the south.[16]

The role of tanks was clearly decisive in the crushing defeat of France. The ledger of German strengths and allied weaknesses should not undermine the fact that tanks had changed the nature of warfare forever. Indeed, as proponents of offense-defense theory would argue, the exceptional tactical and strategic mobility of tanks contributed immensely to the quick and decisive German offensive. On a tactical level, the mobility of tanks allowed for the rapid concentration and breakthrough against enemy weak points, envelopment of defensive strong points by the flanks and rear, and general ability to move and fight in formation while advancing. On the operational level, the mobility of tanks and especially of the motor trucks and tracked infantry vehicles that supported the tank forces was a crucial element of Germany's deep strategic penetration after the Sedan breakthroughs and turning movement to the coast. In short, the groundbreaking speed of the German armored penetration helped produce both the physical and psychological collapse of the Allied defense.

Although the German campaign in France stands as a dramatic example of what tanks and tank tactics could accomplish on the attack, it is easy to imagine a far different outcome if France had been stronger in intelligence and command and control or if the Allies had employed their forces more effectively on defense. In analyzing the French campaign, Liddell Hart argued:

> Never was a world-shaking disaster more easily preventable. The Panzer thrust could have been stopped long before reaching the Channel by a concentrated counterstroke with similar forces. . . . The thrust could have been stopped earlier, on the Meuse if the French had not rushed into Belgium leaving their hinge so weak, or had moved reserves there sooner. . . . The thrust could have been stopped before it even reached the Meuse if the approaches had been well covered with minefields. It could have been stopped even if the mines were lacking—by the simple expedient of felling trees along the forest roads which led to the Meuse.[17]

Liddell Hart's conclusion is perhaps an exaggeration—if simply because it minimizes deeply rooted problems with Allied doctrine, training, and organization—but not by much. It is highly unlikely that Germany could have achieved its stunning offensive in France if the Allies had been more adept at using armored forces. Specifically, the German attack probably would have been stymied if the French had placed some of their better divisions (infantry, artillery, or otherwise) opposite the Ardennes, deployed in greater depth their forces already in the area of the main attack, committed their tanks in a concentrated and coordinated fashion against the point of the breakthrough, or withheld their mobile reserves in the north and instead shifted to hit the vulnerable flanks and logistical tail of the penetrating armored spearheads.[18]

The German invasion of the Soviet Union in June 1941 was initially enormously successful. The tactical and strategic offensive capabilities of the German panzer and motorized divisions were vital to the spectacular breakthroughs, penetrations, and encirclements of entire Soviet armies at Minsk, Smolensk, Vyazma, and Kiev. Once again, a key to German success lay in the glaring weaknesses and blunders of its adversary.[19] The Russians possessed an equal or greater military force in terms of sheer numbers and quality (including at least a four-to-one advantage in tanks), but Stalin's purges of 1937–38 had thoroughly drained the Red Army of its best commanders and officers. As one historian describes the purges:

> Whatever Stalin's motives, and whether or not he intended to go as far as he did, the final figures were staggering. Only [two] survived among the marshals. Out of eighty members of the 1934 Military Soviet only five were left in September 1938. All eleven Deputy Commissars for Defence were eliminated. Every commander of a military district (including replacements of the first "casualties") had been executed by the summer of 1938. Thirteen out of fifteen army commanders, fifty-seven out of eighty-five corps commanders, 110 out of 195 divisional commanders, 220 out of 406 brigade commanders, were executed. But the greatest numerical loss was borne in the Soviet officer corps from the rank of colonel downward and extending to company commander level.[20]

The depleted state of the leadership of the Red Army stood in stark contrast to the highly skilled and combat-experienced commanders of the German army on the eve of the Soviet campaign. Moreover, with the partition of Poland in 1939 and occupation of Bessarabia, Bukovina, and the Baltic states, the bulk of the Red Army had deserted strong defensive positions of the "Stalin line" and moved farther west into more vulnerable dispositions. Stalin compounded the matter by ordering the army, now bunched on the new frontier, to stand and fight in place rather than allowing mobile forces to defend in depth. Last, as with the Allies in the west, the Soviet army scattered its armored forces equally among infantry units, rather than concentrate the tanks in independent units, because tanks were primarily cast in an infantry support role.

These Russian deficiencies notwithstanding, the German invasion of the Soviet Union exhibited the offensive potential of greater strategic mobility conferred by the tank. However, initial German operations also revealed and foreshadowed the significant defensive capabilities of armored forces, as well as important offensive limitations. The German method of encircling and destroying Russian forces at the outset of the invasion included an important defensive element. In these offensive maneuvers the Germans, like the Prussians over seventy years earlier, sought to place their attacking forces in a position from which they could conduct tactical defensive operations.[21] The Germans proved adept at using armored forces in defensive crises as well. In mid-August 1941, for example, the Russians launched a strong counteroffensive south of Lake Ilmen, quickly penetrating into the rear and threatening a German flank and network of communications. Hitler immediately ordered a panzer corps to detach from its army group, redeploy rearward in a wide arc, and counterstrike the flank of the Russian penetration. The Russian offensive quickly collapsed.[22] In short, there is ample evidence even in the early stages of World War II that the benefits of being able to fight mobile armored battles were not limited to the attacker.

Finally, despite the dramatic encirclements of huge numbers of Red Army forces, German operations in the Soviet Union in the early months of the war revealed significant problems with offensive armored warfare. First, Germany struggled to supply its armored forces across the vast expanses of territory in the Ukraine and Byelorussia. The smaller scale of operations against Poland and France had partly masked this crucial logistical problem, but the inability to keep fast-moving, fuel-guzzling German armored forces supplied was manifest. Second, the German army found it difficult to coordinate the movement of tanks and troops. Even independent armored formations relied on slower-moving infantry to achieve, secure, and capitalize on battlefield victories, but the inability to coordinate the actions of the two forces hampered operations from the start of the Soviet campaign. These twin problems marked a fundamental weakness in the entire concept of decisive offensive tank warfare.[23]

Armored Warfare in World War II, 1942–44

In the campaigns of 1939–41, the German army's doctrine and strategy for its tank forces were superior to those of its adversaries. This largely explains the series of quick and decisive victories for the attacker in the period. The evidence from another period of the war—1942 to 1944—does not support labeling the tank as an offense-dominant innovation.

The best evidence with which to evaluate the offense-defense impact of tanks comes from later in World War II when both attackers and defenders had learned the essentially optimal methods for armored warfare. Proponents of offense-defense theory acknowledge the problem of asymmetry in optimal force deployment and doctrine that prevailed at the start of World War II and agree that the appropriate evidence comes from later in the war "when the Allies had also realized the uses of these technological advances and deployed appropriate forces and doctrines on a broadly even footing with the Germans."[24] The most pertinent period is from operations on the eastern front in the winter of 1942–43 through the final German offensive on the western front at the end of 1944. This period is crucial because it marks the time from when the Russians learned to conduct offensive and defensive armored warfare at a reasonably proficient level until the time when any impact of tanks on the offense-defense balance was eclipsed by the sheer imbalance of material power between Germany and its adversaries.

By the winter of 1942–43 the Russians had developed the necessary organizational and doctrinal expertise to conduct offensive and defensive armored warfare at a reasonably proficient level. In terms of defensive warfare, Stalin and his commanders learned from both the catastrophic defeats suffered in the first year of the war and the experiences of the battles at Moscow, Leningrad, and elsewhere that the density of tank forces and depth in defense were the keys to success. The Russians also came to understand that a strategic retreat, such as that undertaken by the Red Army in the south in July 1942, was the best option in the face of a massive German armored offensive.

In terms of offensive warfare, the Soviet experience was more one of rediscovering old lessons. By 1936, the Soviets had in fact developed the most advanced armor doctrine and organization in the world. The centerpiece of Red Army doctrine at that time—the concept of "deep battle"—was nearly identical to contemporaneous Wehrmacht doctrine for employing tanks in warfare. Both German and Soviet officers believed tanks would be most useful if employed in an offensive, mobile, combined-arms style of battle carried deep into the defender's rear in order to disrupt and destroy operational command and logistics. Unfortunately for the Soviet Union, for a variety of reasons (including the execution in 1937 of the leading tank innovators, such as Mikhail Tukhachevski, as part of Stalin's purges) the Red Army was unable to implement the doctrine in practice and had completely discarded the concept of deep battle for a more conservative armored doctrine in the years before the German invasion.[25] Only after Germany quickly crushed Poland in 1939 and France in 1940 did the Red Army start to rethink its armored doctrine. By the summer of 1941, the Soviet Union had only begun to reintroduce the organizational and doctrinal changes consistent with the original deep-battle concept.

The learning process was particularly painful on the attack. In May 1942, for example, the Red Army sought to disrupt German planning for the summer offensives by attempting to counterattack and encircle German forces in the Kharkov region. This first trial of the newly organized Soviet armored units was fraught with mistakes and ended in disaster, with the loss of the equivalent of four Soviet armies. From this experience, however, Russian planners recognized the need to employ large armored forces properly, correctly identifying the problems of committing mobile formations in penetration, exploitation, and pursuit.[26] These lessons bore fruit in November 1942, when the Soviets conducted a brilliant redeployment, used large mobile formations to penetrate Axis defenses north and south of Stalingrad, and encircled almost three hundred thousand German troops. By the time the first modern Soviet tank armies were officially created in January 1943, the Red Army had (re)integrated the necessary knowledge and skill to conduct effective armored operations.

Two factors related to asymmetries in power and skill should be noted, as they appear to make even this period of World War II less than a perfect test case for assessing the impact of tanks. First, the German army was steadily outnumbered and outgunned. On November 1, 1941, Germany and its allies held a 1.9 to 1 advantage in combat forces over the Soviet Union and its allies on the eastern front. On November 1, 1942 the balance had shifted in favor of the Soviets at a ratio of 1.74 to 1. The balance worsened by October 14, 1943 (2.15 to 1) and November 1, 1944 (3.02 to 1).[27] Germany's shortage of manpower and resources prevented it from deploying a true defense in depth against expected Soviet armored attacks, which would have been the optimal strategy. Second, Hitler increasingly interfered in all aspects of military operations, often to the detriment of German fighting power. In his infamous Defense Order of September 8, 1942, Hitler stipulated that "no army group commander or army commander has the right to allow on his own authority the execution of a tactical withdrawal without my specific approval."[28] This draconian no-retreat policy generally prevented the German army from conducting a flexible, maneuver-oriented defense to which it was best suited and which would have been more effective against Soviet armored offensives.

The resource and command constraints on the German force do not undermine the relevance of this period as a good test of offense-defense theory. To foreshadow the argument, the German army was equally as adept at conducting armored warfare on the defensive as on the offensive, if not more so. That the Germans suffered a great imbalance of military forces only strengthens the case that the mobility conferred by tanks could prove as effective in defense as in the attack. Moreover, the view that the tide turned against Germany because of Hitler's erroneous command and the vast numerical superiority of the Soviet army is somewhat exaggerated, especially for this period. From 1943 to 1944, while infantry and other parts of the German army suffered chronic shortages and declined rapidly, the armored and mechanized forces (under the control of Heinz Guderian) were repeatedly rebuilt and kept strong.[29] In addition, the memoirs of German army commanders are quick to blame Hitler for everything that went wrong on the eastern front,[30] but in reality there were often few good strategic or operational alternatives to the course taken by Hitler. More important, although the Soviet Union was ultimately able to bring greater force to bear on the battlefield, the notion that the Red Army was victorious because it was able to employ a "steam-roller" strategy with an inexhaustible supply of manpower is simplistic. The Russians were able to develop their own sophisticated understanding of how to stop armored penetrations and conduct mobile warfare, though steamroller tactics were often employed. Recent histories of military operations in the east during World War II suggest that the Russians' use of encirclement and deep battle were in some ways even superior to anything achieved by the Germans.[31] In short, the Russians in this period became capable of fighting on a roughly equal footing with German armored forces.

The nature of warfare in this crucial period demonstrates that the mobility conferred by tanks did not favor offense. The Russian encirclement of the German Sixth Army at Stalingrad in November 1942 marked the turning point of World War II. The failure of both major German offensives launched in the summer of 1942—one aimed at capturing Stalingrad, the other intended to overrun the Russian oil fields in the Caucasus—can be credited to a combination of fierce Russian resistance, Hitler's faulty strategy, and enormous German logistical problems. In any event, from this time on the Red Army was on the strategic offensive, while the Germans adhered to a fundamentally defensive strategy. Time after time, however, the Germany army relied on the mobility of its armored forces to halt and defeat major Soviet offensives. The series of defensive operations fought by the Germans in southern Russia and the Ukraine in the winter of 1943 are particularly informative because they were conducted largely free of Hitler's rigid defense dictum. In addition, the Soviet and Allied defeat of the largest German offensives in this period, at Kursk and the Ardennes (respectively), suggest that the spectacularly successful armored offensives of 1939–41 were an aberration not to be repeated against opponents skilled at armored warfare.

German army doctrine regarded speed, mobility, and counterattack as the decisive elements of defense. Tank forces thus provided the perfect tool of defense, especially on the eastern front, where the vast space and extended front lines prevented any textbook elastic defense in depth.[32] Instead of wearing down a Soviet armored penetration with a sequence of defensive lines, the Germans discovered early on that an excellent way to defeat an offensive armored breakthrough was by immediate counterattack with tanks against the flanks of the penetrating force. Again, speed and mobility were seen to be far more important than mass and firepower in conducting these counterattacks. Thus, as one panzer general noted, "the armored divisions, originally organized as purely offensive formations, had become [by early 1943] the most effective in defensive operations."[33] Similarly, in studying the problem of defending the western front in 1943, Heinz Guderian noted, "Our opinion was that it all depended on our making ready adequate reserves of panzer and panzergrenadier divisions: these must be stationed far enough inland . . . so that they could be switched easily to the main invasion front once it had been recognised."[34]

German defensive tactics using mobile armored reserves proved invaluable in parrying Soviet counterattacks as early as 1942. In July and August, Soviet armored thrusts penetrated the front lines of German Army Groups Center and North in several places and were only repelled by the rapid movement and counterattack of local armored reserves.[35] In December, the Germans halted a major Soviet offensive against Army Group Center in the Rzhev salient, as several panzer divisions rushed to seal off the Soviet penetrations.[36] Also in December, the Germans held the defensive line of their Stalingrad relief force when a Soviet tank penetration into their rear was destroyed by a panzer division undertaking repeated counterattacks against the Soviet forces' flanks and rear.[37]

The power of a tank-oriented defense was best displayed by Field Marshal Erich von Manstein's operations against major Soviet offensives in southern Russia and Ukraine from January to March 1943.[38] The Russians, following their counteroffensive at Stalingrad, launched a general offensive in early 1943 aimed at producing a total collapse of the German southern wing on the eastern front. One phase of this offensive sought to push the Germans west across the Don basin toward the Dnepr River and cut off and encircle a German panzer army in the Caucasus. Although Hitler's no-withdrawal policy was in effect, Manstein demanded and received greater autonomy and flexibility in command as a condition for accepting the responsibility of halting this offensive. This independence allowed Manstein to conduct his defensive operations with "a measure of flexibility, economy, and fluid maneuver unsurpassed on the Russian Front during the entire war."[39] The results were impressive. Though lacking the forces necessary to fight a true mobile defense, Manstein allowed Soviet penetrations in some sectors, ordered stubborn positional defense in a few other sectors, and rapidly shifted and assembled panzer units for counterattacks against the most threatening breakthroughs. The mobility built into the armored forces was crucial not only for slowing Soviet frontal advances, shifting to parry attempted envelopments, and allowing timely withdrawals but also for conducting rear-guard actions for slower units to disengage and for delivering spoiling attacks on Soviet preparations.

The second phase of the Soviet offensive was aimed at pushing the front west to the cities of Kursk and Kharkov. Manstein again coordinated a similar series of panzer withdrawals, concentrations, counterattacks, and encirclements against the advancing Soviet tank forces. As the Soviet armored spearheads ground down, Manstein was able to assemble a potent armored force to strike at the Soviet flank south of Kharkov and roll back Soviet forces north and south of the city. (This counterstroke, combined with another panzer counterattack on the flanks of a Soviet army's penetration west of Kursk, resulted in the famous Kursk salient.) Manstein's skillful use of armored forces allowed for the successful extrication of the panzer army in the Caucasus and crushing of the Soviet spearheads at Kharkov. Against a numerical balance of seven to one, Manstein stabilized the southern front and prematurely ended the Soviet winter offensives. Mobile armored units proved uniquely suited for these demanding defensive operations.

The Germans were not alone in using the inherent mobility of tank forces to stop enemy armored offensives. The final German strategic offensive of the war on the eastern front aimed at pinching off and destroying the Soviet forces in the Kursk salient in July 1943. This was a far more limited goal than the deep penetrations and multiple encirclements of the earlier German offensives, but Hitler intended to use the same successful formula of massing tanks for a lightning blow against the Red Army. Hitler "intended that the Kursk operation should be a tank battle because he wanted a quick decision based on a double armoured envelopment";[40] he thus committed seventeen panzer divisions (with about three thousand tanks, including the new technologically superior Panther and Tiger tanks) to that end. For their part, the Soviets prepared a formidable defense in depth, as their forces in the Kursk bulge were an obvious target for a German attack. The defensive system included an elaborate network of antitank strong points and regions laced with minefields and antitank guns. The most important part of the Soviet defense was the large strategic armored reserve used to reinforce threatened points and retake lost ground. On July 5, the Germans launched their offensive with armored penetrations on the north and south shoulders of the salient. The spearheads were rapidly worn down by antitank defenses and then crushed by the Soviet armored reserves. The battle of Kursk, the greatest armored battle in history, was indeed a quick one but resulted in a clear and decisive victory for the defender. The defeat was the first time a German offensive had been halted before it could break through enemy defenses into the strategic depths beyond.[41] In short, although antitank weapons and barriers were crucial, the mobility of Soviet tanks provided a key part of the defensive solution to the problem of German armored offensives.

After Kursk, the fighting strength of the German army declined rapidly as it was pushed back along a broad front in an unrelenting series of Soviet offensives featuring great masses of armor and artillery. By 1944, the

Germans fielded roughly 3 million troops (including satellite forces) in the east, with 177 divisions (including 26 panzer divisions), 2,304 tanks and self-propelled guns, 8,037 guns and mortars, and 3,000 aircraft. The Soviets fielded almost 6.5 million men, including 515 division-sized formations (including 35 tank and mechanized division equivalents), 5,800 tanks, 101,400 artillery pieces and mortars, and 13,400 aircraft.[42] In addition to this enormous numerical imbalance in forces, the quality of training and combat effectiveness of the German army constantly eroded while the Red Army became even more adept in battle.

Despite these great quantitative and qualitative disparities, the Germans fought a skillful withdrawal, shuttling their dwindling armored reserves back and forth to threatened points for effective counterattacks on Soviet armored breakthroughs. At the conclusion of the battle of Kursk, Manstein conducted another adept mobile defense until Hitler ordered a linear defense of the Kharkov sector. In August 1943, when the Soviets broke through this defense, Manstein abandoned Kharkov and fell back to the Dnepr using skillful counterattacks to keep the defensive lines intact. Examples of effective German armored counterattacks against advancing Soviet spearheads include Manstein's counterattack to halt a Soviet tank army near Fastov in November, in which several brigades were lopped off and destroyed by German panzers; the November and December counterattacks by German SS and panzer units to contain the bridgehead of Krivoi Rog on the Dnepr; and the halt of the Soviet offensive to liberate Minsk and all of eastern Belorussia in mid-November.[43]

The remainder of the Soviet offensive campaign, from the winter of 1944 through May 1945, was essentially a war of attrition. The combination of the Red Army's methodical offensives based on the close coordination of infantry, artillery, and tanks and Hitler's renewed obsession with holding territory at all costs (even preventing his forces from consolidating and regrouping defensively during the spring thaw) virtually eliminated the role of armored mobility on the eastern front. The war ended with the Russians never having conducted any large strategic encirclement comparable to the German operations of 1939 to 1941.

The war on the western front provides generally less suitable evidence for exploring the offense-defense impact of tanks, primarily because of the Allied preponderance of power. There were, however, some noteworthy instances of attempts to emulate the armored offensives of 1939–41. For example, Operation Cobra (the Allied breakout from Normandy in July and August 1944) saw armored breakthroughs and deep penetrations, but the Allies were unable to encircle the bulk of the German forces, mainly because the tank forces had to operate in close conjunction with supporting infantry, artillery, and air forces to avoid destruction by other tanks and antitank forces. In December 1944 Hitler launched his last major offensive of the war in the Ardennes forest, using a force including ten panzer divisions. The German armored spearheads penetrated deep into the Allied rear before they were halted by skilled armored maneuvers and counterattacks, as well as tactical air assaults, on the flanks of the German bulge. In sum, it was clear that the Allies, too, now knew how to stop armored offensives.

Political Outcomes

Did perceptions of the offensive superiority of the tank contribute tó the outbreak of World War II? Before coming to power in 1933, Adolf Hitler set forth his goals of German hegemony over Europe and eventual world domination.[44] These objectives entailed a highly offensive strategy and required war with Germany's neighbors. The only questions concerned how, where, and when war would occur. On the question of how, Hitler sought a series of short, decisive campaigns against isolated enemies on the continent. Whatever the underlying reasons for this preference (a combination of domestic, material, and personal factors)[45], Hitler's plan to fight short wars fit well with the German army's doctrine of offensive warfare and rapid maneuver. In terms of where, Hitler envisioned winning decisive campaigns in the east before turning west to defeat France. This would be followed by either an alliance with or defeat of Britain, defeat of Russia, and attack on the United States.[46] On the question of when, the answer for Hitler was simple: as soon as possible. Hitler's urgency stemmed above all from his fear of an unfavorable shift in the balance of forces as the Allies reacted to Germany's head start in rearmament. Hitler began expanding Germany's borders in 1938 with a series of diplomatic triumphs but was frustrated that "the Western Powers had made it so difficult for him to begin war as early as 1938. 'We ought to have gone to war in 1938,' [Hitler wrote,] 'September 1938 would have been the most favorable date.' "[47] As it was, Hitler took the decisive step toward world war when he attacked Poland in September 1939.

German aggression was perhaps inevitable given Hitler's unrelenting determination to establish German hegemony on the continent. He believed that this objective ultimately required waging war on his enemies, and his position of supreme political and military authority in Germany allowed him to pursue his goals with little interference. Donald Watt writes, "Hitler willed, wanted, craved war and the destruction wrought by war. . . . Neither firmness nor appeasement, the piling up of more armaments nor the demonstration of more determination would stop him."[48]

Aside from the question of whether Hitler could have been deterred, we may still be able to assess whether his eagerness to take the offensive was influenced by perceptions of the mobility-enhancing potential of tanks.

Offense-defense proponents claim that Hitler was more willing to attack his neighbors when he perceived offense dominance. In particular, proponents argue that the timing of Hitler's decision to attack the Low Countries and France in May 1940—not earlier in the fall and winter of 1939–40—can be explained by his eventual recognition that armored forces, combined with the blitzkrieg doctrine, had greatly strengthened offense. The remainder of this chapter examines Allied perceptions of the impact of tanks and whether Hitler's eagerness to take the offensive was influenced by his or his military leaders' views of the mobility-enhancing potential of tanks.

Allied Perceptions of Armored Warfare

Tanks had little impact on the battlefield in World War I, as previously noted. But the interwar period witnessed a tremendous debate about how tanks should be integrated into the armed forces. In every major European country, military strategists sought to comprehend what the tank could and should do on the battlefield. No consensus emerged on the likely consequences of armored warfare, but only a few experts concluded that tanks would have a revolutionary impact. Regardless, all the major powers produced and incorporated large numbers of tanks into their armed forces.

From the end of World War I through the early 1930s, Britain led the world in the development of armored warfare doctrine, tank design, and experimentation with training and employing armored forces.[49] In the 1920s, a small group of British military thinkers argued that tanks had revolutionized warfare and could restore offensive superiority to the battlefield. The best-known of these "radical" thinkers, Major General J. F. C. Fuller, saw the tank as the most important technological development of World War I, believed that the mobility and firepower of the tank would dominate future battlefields, and argued that properly employed tanks could result in quick and decisive victories for the attacker. In 1919, Fuller wrote, "From its inception the tank offered the armies which adopted it . . . solutions . . . to the problem of 'how to obtain and maintain offensive superiority.' "[50] Shortly thereafter, and largely influenced by Fuller, B. H. Liddell Hart argued that the inherent mobility of tanks had the potential to revolutionize warfare by shifting the offense-defense balance in favor of offense. British experiments with armored warfare between 1926 and 1934, however, never lived up to expectations, and the mainstream of British military thought saw no great promise for the future of the tank. At best, the British general staff viewed tanks as fulfilling the traditional role of cavalry, that is, in the exploitation of tactical victories won by the infantry and artillery.[51] By the mid-1930s, even the progressive Fuller and Liddell Hart had lost their enthusiasm for tanks as revolutionary offensive weapons and changed their views accordingly. Whereas Fuller became more and more uncertain that an armored offensive could break through a well-organized antitank defense, Liddell Hart completely reversed himself by adopting the argument that defense would dominate in armored warfare.[52]

The French never viewed tanks as revolutionary. To be sure, France devoted considerable resources in the interwar period to building large armored forces and French tanks had greater firepower, better armor, and were more reliable than German tanks.[53] However, French doctrine emphasized tight control over battlefield operations and maneuver and held artillery to be the dominant weapon on the battlefield. Tanks were meant to be used to support the "methodical battle"—a controlled method of warfare that called for the close coordination of infantry, artillery, and tanks, and eventually close air support in an advancing wave of firepower.[54] France did not perceive that tanks gave the attacker the means to a quick and decisive victory.

German Perceptions of Armored Warfare

From the end of World War I through the attack on Poland in September 1939, few (if any) German leaders believed that tanks had fundamentally transformed military operations or strategy. The nature and role of German perceptions of armored warfare at the outbreak of World War II can be understood only in the context of the evolution of German army doctrine and strategy in the interwar period. This is a complex subject that has generated an enormous amount of historical analysis and debate.[55] For present purposes, two aspects of the evolution of German strategy and doctrine are noteworthy. First, the German army was already committed to mobile and offensive warfare before the emergence of tanks and rise of Hitler. Second, and contrary to popular belief, the evidence suggests that neither Hitler nor his top military commanders adopted a radically new doctrine (only later labeled "blitzkrieg") based on perceptions of a great offensive advantage of armored forces. Recent historical scholarship questions the degree to which Hitler and the Wehrmacht *ever* adopted such a revolutionary armored doctrine. Regardless, the evidence shows that nothing of the sort guided Hitler's thinking when he decided to attack France in late 1939, months before any plausibly new blitzkrieg strategy was proposed. In short, it is difficult to find evidence that German perceptions of armored offense dominance contributed to the outbreak of World War II.

In the 1920s, the German army under the leadership of General Hans von Seeckt returned to its traditional military doctrine of seeking quick and decisive victories through highly mobile offensive warfare. This commitment drew on a rich tradition extending back to Helmuth von Moltke (the elder) and Alfred von Schlieffen, if not to the reign of Frederick the Great. The historian Matthew Cooper summarizes the principles of this tradition of mobile offensive warfare: "The quick concentration of force at the decisive point was indispensable, the grand sweeping movements of encirclement basic, and the total destruction of the enemy paramount. Victory was seen to lie in strategic surprise, in the concentration of force at the decisive point, and in fast, far-reaching concentric encircling movements, all of which aimed at creating the decisive *Kesselschlachten* (cauldron battles) to surround, kill, and capture the opposing army in as short a time as possible."[56]

The German army regulations that appeared in 1921 and 1923 reflected Seeckt's reforms and emphasized the importance of fast and decisive maneuver, an offensive mind-set, decentralized command, and the exercise of quick judgment and individual initiative on the battlefield.[57] There were multiple reasons for the return to a thoroughly offensive strategy: First, reliance on a well-established military tradition provided organizational stability and certainty in an unstable and uncertain world.[58] Second, Germany required the capability for quick offensive action and rapid maneuver to sequentially defeat its enemies and avoid fighting a prolonged war on several fronts.[59] Finally, the Treaty of Versailles—which prohibited Germany from forming a mass army, building fixed fortifications in the west, and having heavy forces of all types—compelled the German army to emphasize rapid offensive maneuver as a means of defeating more powerful armies before they could mobilize.[60] Whatever the underlying causes, it is clear that the Germany army had reverted to its traditional strategic preference for quick and decisive victories based on mobility and the offensive well before the rise of Hitler and Nazi Germany.

Despite this emphasis on restoring offensive mobility to the battlefield, most Germans discounted the offensive combat potential of tanks. The experience of World War I had shown that tanks could have a "moral effect" on unprepared troops but could easily be defeated by defensive countermeasures. "In the German assessment, tanks were similar to poison gas and flame-throwers as technological nuisances without decisive potential."[61] The Germans observed and reported on the British armored experiments in the late 1920s but did not produce any sophisticated literature on tank warfare in that period.[62]

By 1930 the Germans had developed and tested a few prototype tanks despite the Versailles restrictions. While many top officers remained skeptical of the utility of tanks, most officers had come to believe that tanks would have an important role to play in any future war. Moreover, armor enthusiasts generally viewed the tank as more of an offensive weapon than a defensive one. However, the dominant view was that the primary contribution of tanks would be in supporting and escorting the regular infantry. In fact, Seeckt still advocated considerable reliance on horse cavalry over tanks for decisive offensive maneuver.[63] Little about German tank doctrine had changed by the mid-1930s: typically, tanks would be used in close cooperation with the infantry to create breakthroughs or achieve envelopments of enemy forces.

The myth persists that during the late 1930s the German army developed a new doctrine of warfare—the blitzkrieg—based on the revolutionary potential of armored forces to achieve a quick and decisive victory for the attacker, which they then employed with great success in 1939–41.[64] One of the key figures in the conventional blitzkrieg narrative is Heinz Guderian, a captain and then general in the German army. Guderian strongly believed that tank forces should not be tied to slower-moving infantry; instead, he (along with several others) advocated the use of independent armored formations to break through the enemy's front and conduct deep strategic penetrations. Whether or not Guderian deserves most of the credit for developing these ideas remains controversial, although he clearly played a major role in the implementation of German armored doctrine before the war. However, Guderian's ideas were met with skepticism, resistance, and sometimes outright subversion by the senior leaders of the army throughout the 1930s.[65] High-level military discussions in 1938 and exercises in the summer of 1939, immediately before the Polish campaign, led German leaders to fear that a future war involving large numbers of tanks would still devolve into the positional stalemate of World War I.[66] This kind of thinking is inconsistent with the traditional depiction of the German army's discovery of a war-winning method of armored operations.

Guderian did not view his own theory of the future of armored warfare as radically new. He emphasized that ultimate success on the battlefield could not be achieved by armored units acting alone.[67] Other historians question whether Guderian's ideas were anything more than a restatement of the traditional German strategy of seeking envelopment and decisive battle.[68] This conclusion is supported by the fact that the German attack on Poland was based not on a blitzkrieg strategy but on the traditional German strategy of a combined-arms attack on the flanks in search of a decisive envelopment of enemy forces. In sum, Hitler eagerly attacked Poland despite the absence of any new revelations of the offensive power of tanks.

To summarize, from the end of World War I through the attack on Poland, few German military leaders were convinced that the tank had revolutionized warfare by restoring offensive superiority to the battlefield. The strategic and operational significance of independent armored forces was not even clear to Hitler or the bulk of the German army before 1940. Because Germany's traditional military doctrine was highly offensive and mobility-oriented, however, the army was better able than its adversaries to develop and incorporate armored forces.

The period from the end of the Polish campaign to the attack on France in May 1940 presents a crucial case for offense-defense theory. Here the best argument can be made for the proposition that Hitler's perception of the offensive potential of the tank made him more willing to initiate military conflict. Specifically, the standard argument is that Hitler was deterred from invading France in the fall of 1939 and winter of 1940 because he lacked a plausible theory of victory. He attacked in May 1940 only after the blitzkrieg, with its promise of employing tanks in a unique way to achieve a rapid and decisive victory over the Allies, became a viable option.[69]

Hitler was not deterred from attacking France in late 1939 and early 1940, and he planned to do so without any new model for employing tanks.[70] At the conclusion of the Polish campaign, Hitler met with his military commanders and announced that he had decided to attack in the west as soon as possible: "The sooner, the better," Hitler said, "Do not wait for the enemy to come to us, but rather immediately take the offensive ourselves."[71] The operational plan drawn up by the German army and endorsed by Hitler in October 1939 called for an attack through Holland, Belgium, and Luxembourg to defeat as many of the French and Allied forces as possible and to capture a large portion of the English Channel coast for subsequent operations against Britain and the remainder of French territory. Hitler was completely confident in a German victory, as he was with every war plan drawn up through May 1940, but he realized that this plan could achieve only a limited territorial objective and would probably lead to a war of attrition. Most important, Hitler did not believe that tank forces offered the potential for a decisive victory: "Hitler had not yet fully grasped the revolutionary potential of the tank. Specifically, no evidence indicates a belief on his part during this period that panzer divisions could be employed so as to produce the decisive victory in the west that he wanted so badly."[72]

All of Hitler's top military commanders believed that the initial war plan against France would almost certainly lead to a catastrophic defeat for Germany. The army chief of staff, General Franz Halder, and the commander in chief of the army, General Walther von Brauchitsch, argued that German armored formations moving into Belgium and France would be destroyed by Allied antitank fortifications, bombers, artillery, and superior tank forces. The German field commanders, including those in charge of huge armored forces, concurred that an offensive against France would be disastrous.[73]

Despite his own reservations about the war-winning potential of tanks and the determined opposition of the German military leadership, Hitler fully approved a plan to attack France on October 22, 1939. He set November 12 as the date for the beginning of the offensive. Hitler dismissed all arguments for postponing the attack presented by Brauchitsch and Halder and unequivocally reiterated the November 12 launch date. The attack was postponed several times in November by poor weather conditions, as were a series of rescheduled offensives through December. On January 10, Hitler received a promising weather forecast and thus fixed an attack for January 17, 1940. General Fedor von Bock, commander of German Army Group B, wrote in his diary, "After the eleventh postponement (the first on Nov. 11, 1939), it appears now as if the operation is finally going to proceed!"[74] On January 10, however, a German plane carrying secret documents relating to the offensive was forced to land in Belgium, thus compromising German intentions. The unlikely sequence of weather delays, the plane crash, and the onset of winter forced Hitler to postpone his attack until the spring of 1940.[75]

The winter delay and the compromised war plan made drastic operational changes both possible and necessary. Hitler eventually accepted a daring alternative plan to feign an attack into Belgium and the Netherlands while concentrating panzer forces for a surprise attack through the Ardennes Forest. The tank divisions would then conduct a deep strategic penetration into the French rear, cutting from the Meuse River to the English Channel coast.[76] Even with the final plan as launched on May 10, 1940, no German military leader expected success. Halder put the odds of victory at ten to one against and wrote to his wife that he and his colleagues thought what they were doing was "crazy and reckless."[77]

In sum, the evidence of German decision making before World War II shows that the political decision to initiate military conflict preceded any perception of the great offensive potential of tank technology. Adolf

Hitler was bent on military conquest from the outset and decided to launch an offensive in the west well before February 1940, when the bold plan to concentrate panzer divisions for a surprise attack through the Ardennes was accepted. Even then the attack was seen as a tremendous gamble, as evidenced by the fact that Hitler and others frequently labeled the victory "a miracle." For Hitler, the gamble was worth taking in pursuit of his political ambitions; for German military leaders, it was simply the only viable military option available under the circumstances.

Conclusion

The introduction of tanks into the European armed forces in the interwar period resulted in an enormous increase in the tactical, operational, and strategic mobility of land forces. The tank's combination of technological advances in the combustion engine, firepower, communications, and armored protection made it the dominant fighting vehicle on the battlefield. According to offense-defense theory, great improvements in mobility shift the offense-defense balance in favor of offense. Offensive shifts in the balance, in turn, result in quicker and more decisive victories for the attacker. Perceptions of offensive advantage make states less secure and more willing to initiate conflict as a means of pursuing their interests.

The evidence presented in this chapter does not lend support to offense-defense hypotheses. In terms of military outcomes, the historical record does not demonstrate that armored mobility gave attackers a decisive advantage over defenders. In World War I and the interwar period, tanks had little effect on operational outcomes. This evidence is not surprising given that military experts had not yet understood the value of tank forces or fully grasped the best methods for employing tanks on the battlefield. In World War II, armored mobility was perhaps the crucial element of Germany's stunningly swift offensive operations in Poland (in September 1939), France (in May and June 1940), and the Soviet Union (in the summer of 1941). These campaigns, however, tell us little about the impact of tanks on the offense-defense balance because of large asymmetries between Germany and her adversaries in power (as with Poland) and skill and doctrine (as with France and the Soviet Union). The more relevant evidence from World War II comes from the period when all sides were relatively adept at armored warfare and had integrated large numbers of armored tank units into their forces and doctrines, and before the Germany army became a spent force. The operational use of tanks in this period of combat—from late 1942 to late 1944—shows that mobile armor was no less valuable for defending forces than it was for attackers.

In terms of political outcomes, few political or military leaders before the outbreak of World War II believed that tanks had given the attacker a decisive advantage on the battlefield. Adolf Hitler's view of armored forces was less significant as a cause of the war than his racist ideology and, especially, aggressive geopolitical ambitions. Although Hitler and the German army understood that the tank had improved military mobility, and they revered mobility for its contribution to offensive warfare, it was not perceptions of armored offense dominance that prompted Hitler to seek his political objectives through military conquest. Once again, politics is the master, technology the servant.

5

The Nuclear Revolution

The nuclear revolution produced a "fantastically great" increase in destructive power.[1] Whereas the most powerful bombs used in World War II contained ten tons of the standard chemical explosive TNT, the average explosive yield of the atomic bombs dropped on Hiroshima and Nagasaki in August 1945 was equivalent to 18,000 tons. The first thermonuclear test, which occurred in November 1952, had a yield of 10.4 million tons, almost 580 times the power of the first atomic bombs and over a million times more powerful than the biggest bombs of World War II. In fact, all of the explosive power used in World War II could fit in one warhead on a modern intercontinental ballistic missile. In short, nuclear weapons produced an exponential increase in firepower beyond historical comparison.

Offense-defense theorists contend that nuclear weapons created a revolution in defensive advantage. That argument is consistent with the firepower-favors-defense hypothesis, but—as with many other strategic concepts in the nuclear age—it is also counterintuitive. Nuclear defense dominance seems counterintuitive not only because no country has ever deployed an effective defense against a well-armed nuclear attacker but also because most people find little security and comfort in either the memory of the cold war "balance of terror" or the prospect of nuclear proliferation.[2] Moreover, unlike with the other cases of technological change examined in this book, we (fortunately) cannot look to historical evidence of nuclear combat to assess the impact of the nuclear revolution on the objective offense-defense balance. Thus, the initial section below on military outcomes is devoted to examining and evaluating the logic of the claim that nuclear weapons are defense-dominant.

Offense-defense theory yields more concrete and testable predictions about the political consequences of the nuclear revolution. Indeed, offense-defense theory emerged from the cold war-era literature on nuclear strategy and deterrence theory, advancing the core idea that the nuclear revolution had rendered strategic nuclear competition between the United States and the Soviet Union irrelevant, irrational, and dangerous.[3] Relying on the idea that the nuclear revolution dramatically shifted the offense-defense balance toward defense dominance and thus made nuclear great powers enormously secure, offense-defense proponents have offered at least three specific predictions about political outcomes. First, war between nuclear states should not occur. Second, security competition between nuclear armed states over distant territory should not be intense. Third, intense arms racing beyond levels of mutual assured destruction (MAD) should be absent. The section on political outcomes below discusses the logic of each of these predictions and compares them to the empirical record from the nuclear era.

How does offense-defense theory fare in this case study? The claim that nuclear weapons give sufficiently armed defenders a large advantage over attackers is difficult to sustain in purely military terms. In fact, as is shown below, the argument that nuclear weapons shift the objective offense-defense balance toward defenders primarily turns on the effects of subjective political perceptions. If restricted to military operational criteria, nuclear weapons could justifiably be categorized as offense-dominant. This categorization, however, would pose an enormous challenge for the conceptual coherence and empirical validity of offense-defense theory.

The empirical evidence from the nuclear era is equally problematic for offense-defense predictions of political outcomes. Nuclear weapons, according to offense-defense theory, essentially eliminate the security dilemma among well-armed states and thus much of the grounds for security competition. Although the nuclear revolution has coincided with a prolonged era of major power peace, the nature and intensity of the nuclear arms race between the superpowers during the cold war undermine offense-defense theory. Proponents are aware, of course, that the record of superpower behavior during the cold war does not mesh with their key predictions. After all, if American and Soviet leaders had fully accepted and behaved according to the dictates of the thesis that the nuclear revolution rendered arms racing beyond the requirements of MAD unnecessary, then offense-defense proponents would not have needed to argue so often and so vociferously about the wrongheadedness of the strategic nuclear competition.[4] Proponents typically offer a number of arguments in defense of their theory. One perspective holds that although the superpowers' procurement and targeting policies suggest an enduring competition for nuclear advantage, when it came to crisis bargaining and diplomacy leaders acted in ways that showed they understood the relative unimportance of the nuclear balance beyond MAD levels. Another line of argument would hold that the apparently suboptimal behavior of the superpowers is explained by American and Soviet leaders' gross misperceptions of the nature of the nuclear revolution and woeful ignorance of its military and political consequences. Thus, in addition to presenting the empirical record, the discussion below explores the issue of whether U.S. nuclear policy during the cold war is better understood as an essentially misguided response to the pacifying effects of the nuclear revolution or a reasonably rational reaction to the uncertainty of technological change in a fundamentally insecure world. In this respect, the nuclear case provides a good opportunity to more directly compare the explanatory power of both offense-defense theory and technological opportunism.

Military Outcomes

Offense-defense theorists contend that the nuclear revolution shifted the offense-defense balance strongly in favor of defenders. "After 1945 thermonuclear weapons restored the power of the defense, this time giving it an overwhelming advantage," Stephen Van Evera writes.[5] According to Charles Glaser, "Nuclear weapons created a revolution for defense advantage."[6] Many traditional military concepts are not easily transferred from the conventional level to the nuclear realm, so it is no surprise that the logic behind this classification of the effects of nuclear weapons on attack and defense is somewhat complicated. Whereas the standard firepower-favors-defense hypothesis rests on the ability to hinder advancing attacker forces, nuclear defense dominance rests on the ability not to stop attacking forces but, rather, to retaliate in kind.

In a strictly technical military sense, the introduction of nuclear weapons created a situation of overwhelming *offense* dominance. For the first time in the history of warfare, prevailing technology (i.e., nuclear warheads delivered by bombers or married to long-range missiles) gave attackers the ability to destroy an adversary quickly and decisively before engaging and defeating that adversary on the battlefield. The change was not due solely to the awesomely greater level of destructive power produced by nuclear weapons than by previous firepower weapons. For example, the American incendiary bombing campaign of Japanese cities that began in March 1945 with the firebombing of Tokyo eventually killed almost one million Japanese—almost ten times the number of civilians killed by the atomic bombing of Hiroshima and Nagasaki in August 1945. As Thomas Schelling notes, "Against defenseless people there is not much that nuclear weapons can do that cannot be done with an ice pick."[7] What really matters in the nuclear age is the speed with which total annihilation can be carried out and the ability to do so without first achieving success on the battlefield. Today, more than half a century into the nuclear era, no state has been able to devise effective defenses against nuclear attack. It thus seems reasonable to conclude that the nuclear revolution dramatically shifted the offense-defense balance in favor of offense. This conclusion is bolstered by the fact that the only instance in which nuclear weapons have been used in war (the atomic bombings of Japan by the United States) occurred as part of an offensive strategy that compelled complete surrender from the adversary.

Offense-defense theorists contend, however, that nuclear weapons are defense-dominant. The key to the argument lies in the dynamics of a situation in which potential adversaries possess secure second-strike nuclear capabilities—that is, when no side can launch a nuclear attack that is devastating enough to prevent nuclear retaliation from the other. Under these conditions of mutual assured destruction, nuclear weapons are seen as defense-dominant because states are deterred from attacking one another. In the nuclear world, deterrence based on the threat of retaliation is the functional equivalent of defense. Robert Jervis writes, "Concerning nuclear weapons, it is generally agreed that defense is impossible—a triumph not of the offense, but of deterrence. Attack makes no sense, not because it can be beaten off, but because the attacker will be destroyed in turn . . . the result is the equivalent of the primacy of the defense."[8]

This formulation appears to make sense because the essential aim of offense-defense theory is to explain the material conditions under which states feel secure and when they do not or, alternatively, when they can defend against or deter attacks and when they cannot. When states rely on deterrence for their security, forces that bolster a state's ability to retaliate overwhelmingly are essentially defensive. In a world of conventional arms, deterrence becomes easier as the defender is increasingly capable of denying territorial gains to the attacker. In the nuclear world, deterrence rests on the defender's ability to punish the attacker with unacceptable costs for attempted aggression. The only way to attack or take territory at an acceptable cost among nuclear armed adversaries would be by eliminating a defender's second-strike capability. This is very difficult to do, offense-defense proponents argue, because the technological characteristics of nuclear offensive and defensive systems make it much easier (less costly) to enhance one's own deterrent forces than to strengthen forces that threaten an adversary's deterrent forces. To illustrate the point, consider the cost ratio definition of the offense-defense balance: the amount of resources that an attacker must invest in offensive forces to offset the amount of resources a defender has invested in defensive forces. Thought of in these terms, and reflecting the strange logic of nuclear deterrence, nuclear weapons would appear to shift the ratio in favor of defense. Even great technological improvements, doctrinal changes, or armament increases aimed at achieving an offensive first-strike capability or limiting damage from nuclear retaliation can usually be readily (and inexpensively) defeated or undermined by protecting and bolstering existing retaliatory forces or by building additional forces.[9] This logic underlies contemporary arguments that U.S. plans for a missile-defense system against even limited nuclear attacks can be easily defeated by simple countermeasures.[10] Strictly speaking, this means that an offensive nuclear strategy dominates a defensive one. However, in terms of deterrence, nuclear weapons strongly favor the defense by giving defenders the ability to inflict unacceptable costs on attackers. As one proponent writes, "The nuclear revolution gave defenders a large military advantage—so large that conquest among great powers became virtually impossible."[11]

One significant objection to classifying nuclear weapons as defense-dominant is worth discussing here. (An additional objection is discussed below in the section on political outcomes.) The argument that the triumph of deterrence is the functional equivalent of the primacy of defense depends on the role of political perceptions, not just objective military interactions. This is a problem as it conflates the two separate logics upon which offense-defense theory rests. According to the theory, at any given time there is an objective offense-defense balance that shapes combat outcomes and a perceived offense-defense balance that shapes political decisions. Coding the objective effects of nuclear weapons on the basis of the political perceptions inherent in deterrence calculations leads to conceptual confusion.[12] After all, by the same problematic logic, the small arms and artillery revolution before World War I, which proponents classify as *defense-dominant,* would need to be recoded as *offense-dominant* because leaders at the time mistakenly (proponents argue) believed that these technologies favored attackers.

Leaving aside this important conceptual problem, the claim that nuclear weapons shift the offense-defense balance toward defenders is reasonably sound. This supports the proposition that improvements in firepower favor defense. The more significant question concerns how the existence of nuclear weapons affects international politics and state security behavior. Do state leaders perceive that nuclear weapons give defenders a large military advantage? If so, do states behave in ways that conform to offense-defense predictions?

Political Outcomes

The consequences of nuclear warfare in a MAD world, according to offense-defense theory, are easy to comprehend and extremely difficult to change. A dramatic and unprecedented shift in the offense-defense balance in favor of the defender brought about by the nuclear revolution should have significant ameliorating effects on the level of international security competition and war. Specifically, offense-defense theory predicts that the incidence of war among nuclear powers should be drastically reduced, security competition over distant territory should not be intense, and arms racing beyond robust MAD levels should not occur.

War

First and foremost, according to offense-defense theory, the prospect of devastation in a nuclear conflict deters even the most highly expansionist states from launching wars of conquest against other nuclear states, and the robust security provided by nuclear weapons virtually eliminates the kind of fear that leads status quo states to launch preventive or preemptive wars. "The first implication of the nuclear revolution is that military victory is not possible," writes Jervis. "From this it follows that if statesmen are sensible, wars among great powers should not occur."[13] The implausibility of military conquest—not to mention a quick and decisive victory—should not only make war among the nuclear powers virtually obsolete but also eliminate much security competition.

An important alternative prediction about the consequences of the nuclear revolution for the likelihood of war is found in Glenn Snyder's "stability-instability paradox." Snyder suggested that stability in the strategic nuclear realm may actually encourage violent conflict at the conventional level. The logic is as follows. If mutual destruction results from nuclear war, then it makes no sense for any state to ever use nuclear weapons. But this stability makes it relatively safe to initiate limited military aggression because the defender's threat to respond with nuclear retaliation is not credible.[14] Proponents of offense-defense theory respond that the stability-instability paradox is flawed as an explanation for international behavior because the likelihood (or even the possibility) of escalation from limited war to nuclear war is enough to deter such provocations.[15] In sum, offense-defense theory predicts no war between nuclear powers given the tremendous defensive advantage of nuclear weapons.

War among the major powers has not occurred since the introduction of nuclear weapons. Numerous serious crises arose between the United States and the Soviet Union during the cold war, but the superpowers never directly engaged in war. Indeed, it is simply hard to imagine any state armed with nuclear weapons and advanced delivery systems being attacked and conquered in the traditional sense. This evidence of congruence between predicted and actual outcomes bolsters offense-defense theory.

The lack of military conflict between nuclear powers, however, needs to be qualified in two ways. First, alternative explanations for the "long peace" of the cold war era abound: many scholars argue that bipolarity, economic integration, and political and normative changes were more responsible for preventing the cold war from becoming hot.[16] Second, history reveals important cases that cut against the no-war logic of the nuclear revolution. For example, a state armed with nuclear weapons has been attacked (Israel via ground invasion in 1973 and missile attacks in 1991), a state has intervened in a war against a nuclear power (China in Korea against the United States in 1950), and two states possessing nuclear weapons have fought each other (the major armed clashes between China and the Soviet Union in 1969).[17] Perhaps most significant, the military conflict between Pakistan and India, both nuclear powers, in the Kargil mountains of Kashmir in the spring of 1999 was a war by traditional casualty measures.[18]

The relevance of these qualifications can be debated. On the one hand, one might argue that alternative explanations for major power peace are flawed or can be subsumed under the nuclear peace explanation and that the possible historical exceptions to the hypothesis are either nothing more than close calls or individual anomalies that do not offer powerful disconfirming evidence for offense-defense theory. On the other hand, critics might respond that because the nuclear era has been relatively short, the finding of nuclear peace is not robust and should be highly sensitive to even a single contradictory finding. In sum, it seems fair to conclude that the horrendous costs likely to be associated with nuclear war have played a major role in preventing the outbreak of general war between the Soviet Union and the United States during the cold war.

Third World Interventions

A second explicit prediction of offense-defense theory about behavior under nuclear defense dominance is that states should not compete or fight too intensely over territory beyond the homeland or the homeland of close allies. Nuclear weapons devalue traditional concerns over geographic depth; in other words, buffer zones and distant bases are less important in a nuclear world because even without them nuclear retaliation can be assured.[19] In the cold war, offense-defense theory would predict minimal intervention and competition between the superpowers in the third world. As Barry Posen and Stephen Van Evera wrote at the time,

> Nuclear weapons make conquest much harder, and vastly enhance the self-defense capabilities of the superpowers. This should allow the superpowers to take a more relaxed attitude toward events in third areas, including the third world, since it now requires much more cataclysmic events to shake their defensive capabilities. Whatever had been the strategic importance of the third world in a nonnuclear world, nuclear weapons have vastly reduced it.[20]

One need not investigate the historical record for long to conclude, as offense-defense proponents clearly did, that the superpowers took anything but a "relaxed attitude" to competition and intervention in the third world.[21] In fact, most crises between the United States and the Soviet Union occurred over clashes of interest in the third world. In these regions, each superpower resorted to the whole range of economic, political, and military means to advance its own interests or block the influence of its rival.[22] Particularly confounding for offense-defense theory is the fact that the Soviet achievement of nuclear parity with the United States, which should have mitigated Soviet security fears, coincided with an *increased* level of intervention in the third world. Indeed, some argue that the Soviets were constrained from too overtly challenging the United States in the third world early in the cold war because of the U.S. possession of nuclear superiority. The arrival of strategic nuclear parity, however, coincided with a much more assertive role for the Soviets, as demonstrated by their actions in the Middle East (1970–73), Angola (1975–76), Ethiopia (1977–78), Yemen (1978–79), and Afghanistan (1979). The emergence of nuclear parity did not mitigate, and may have in fact aggravated, the superpower competition for influence in the third world.[23] Of course, the United States intervened with its own military forces in the third world throughout the cold war as well, in Korea (1950–53), Egypt (1956), Lebanon (1958 and 1982), Thailand (1962), Laos (1962–75),

Vietnam (1964–73), Congo (1964 and 1967), the Dominican Republic (1965), Cambodia (1970), Libya (1981 and 1986), Grenada (1983), and Panama (1989).[24]

How detrimental is this evidence for offense-defense theory? Proponents could offer several counterarguments for why offense-defense theory is not undermined by the evidence of third world interventions. First, proponents could argue, paradoxically, that the superpowers' possession of mutual second-strike capability created space for such interventions because nuclear weapons were not a credible deterrent to them. For example, the United States could not credibly threaten to use nuclear weapons (and thus invite retaliation) if the Soviet Union intervened in Angola or Afghanistan. Likewise, the Soviet arsenal could do nothing to prevent the United States from intervening in Vietnam. However, the logic of offense-defense theory implies that these interventions need not occur in the first place, as they could hardly be necessary for protecting national security. Second, offense-defense theorists could argue that the superpowers had other security interests beyond defense of the homeland, as well as nonsecurity interests, which may have compelled a certain level of intervention and competition abroad. Indeed, the very fact of nuclear weapons possession may have given the superpowers an added measure of independence, which they felt free to use in pursuit of nonvital or nonsecurity interests.[25] However, the fact that the superpowers were so consistently drawn into real and frequently costly conflicts in areas of little economic or strategic value tends to undermine this counterargument. Third, Van Evera concedes that "the two superpowers competed far harder—in both Central Europe and the third world—than objective conditions warranted," but he argues that this behavior resulted from misperceptions of the implications of the nuclear revolution.[26] This and other nonstructural explanations (e.g., domestic politics, bureaucratic politics, ideological imperatives) are plausible, but it seems at least equally (and likely more) plausible that the observed frequency of interventions reflected the strategic political ambitions, relative gains concerns, and other security imperatives inherent in a world of competing great powers, which are not easily mollified by perceptions of technological defense dominance. Although not a defense of offense-defense predictions in this case, a final explanation for superpower intervention in the third world during the cold war is the idea that reputational considerations drove the observed behavior. That is, what was at stake in this kind of competition in the third world was the superpowers' image of each other; thus, interventions were a test of wills and method of heading off future challenges to each other's power.[27] In any event, the record of superpower intervention in the periphery during the cold war is inconsistent with the offense-defense prediction for security behavior in a nuclear, defense-dominant world.

Arms Racing

"The neglected truth about the present strategic arms race between the United States and the Soviet Union," McGeorge Bundy wrote in 1969, "is that in terms of international political behavior that race has now become almost completely irrelevant. The new weapons systems which are being developed will provide neither protection nor opportunity in any serious political sense. Politically, the strategic nuclear arms race is in a stalemate."[28] Offense-defense theory predicts that arms racing should not occur once states believe they have acquired the capability for assured nuclear retaliation. Jervis explains the logic:

> If national security is provided by one's capability to destroy the opponent, not by the possession of a more effective military machine than the other side, then the force that drives the security dilemma is sapped. The security dilemma is created by the fact that in the prenuclear era weapons and policies that made one country secure made others insecure. An army large enough to protect the state was usually large enough to threaten a neighbor with invasion, even if the state did not intend such a threat. But when security comes from the absolute capability to annihilate one's enemy, then each side can gain it simultaneously. Neither side need acquire more than a second-strike capability and, if either does, the other need not respond since its security is not threatened.[29]

By this logic, the United States and the Soviet Union should have felt tremendously secure during the cold war. The United States possessed a secure second-strike capability, which meant that it should have been easy for the country to protect itself from a Soviet attack through deterrence. Because the Soviet Union also possessed a nuclear retaliatory capability, each side could destroy the other in retaliation for an attack. In other words, MAD reigned and the security dilemma should have been virtually eliminated. Charles Glaser sets forth the consequences:

> The superpowers' deployment of large survivable nuclear arsenals established clear defense-dominance, and the technology of nuclear weapon delivery systems and various types of offensive counterforce provided the opportunity to distinguish offense and defense. . . . Under these conditions, [offense-defense theory] predicts a major role for arms control or other non-competitive policies. *The nuclear arms race should have ground to*

a halt and the full spectrum of the most threatening nuclear forces should have been limited either by arms control agreements or unilaterally.[30]

The basic offense-defense prediction about arms racing contains both a quantitative and qualitative element. In quantitative terms, according to offense-defense theory, adversaries will not be too concerned with comparing the relative size of their nuclear arsenals because even large shifts in relative force levels pose little threat to the "weaker" side's ability to retaliate and inflict unacceptable damage. "If the defense is much more potent than the offense," Jervis writes, "each side can be willing to have forces much smaller than the other's, and can be indifferent to a wide range of the other's defense policies."[31] In qualitative terms, offense-defense theory predicts that once states find themselves in a MAD world, they should not attempt to gain an advantage at the nuclear level by building offensive counterforce weapons (i.e., forces aimed at destroying an adversary's strategic nuclear weapons). The logic underlying this prediction is that possession of an assured destruction capability already provides states with a high degree of security; a first-strike advantage is virtually unattainable, impossible to maintain, and thus irrational to pursue.

The evidence from the cold war shows, instead, that U.S. and Soviet nuclear policies were highly competitive and offensive-oriented. The superpowers' nuclear arsenals—in both quantitative and qualitative terms—far exceeded the capabilities necessary for assured retaliation and exhibited a far greater sense of insecurity than can be accounted for in offense-defense theory.

The remainder of the chapter not only presents the relevant evidence from the cold war nuclear arms race, which contradicts offense-defense reasoning, but also attempts to show that technological opportunism is more consistent with the record. To be sure, other nonrealist (and nonmaterial based) arguments would need to be brought into the analysis to provide a fuller picture of the complexity of U.S. nuclear strategy, doctrine, and deployments during the cold war.[32] The discussion also cannot analyze the policymaking process in enough historical depth here to show what arguments were decisive in bringing about policy decisions and implementation. In short, the following discussion does not demonstrate causality and should not be considered a definitive test of technological opportunism. Caveats aside, however, given that offense-defense scholars (and many others) have been so consistently puzzled by U.S. behavior in this case, even a preliminary probe of the validity of technological opportunism seems in order.

The evidence suggests that American leaders were keenly interested in preserving their monopoly on nuclear weapons from the earliest days of the cold war in the 1940s, considered preventive war and actively planned for preemptive war against the Soviet Union and China until the early 1960s, and remained deeply interested in transcending MAD and establishing nuclear superiority throughout the remainder of the cold war. The empirical record is discussed in two parts. First, U.S. nuclear policy from 1945 to the early 1960s is examined. Although this period falls outside the bounds of offense-defense predictions, which come into play after the superpowers acquire mutual assured destruction capabilities, it is significant because it indicates the extent to which American (and Soviet) leaders appreciated the military and political significance of U.S. nuclear superiority. The second part of the section examines evidence from the early 1960s onward, during the period in which both the United States and the Soviet Union possessed robust retaliatory capabilities but pursued much first-strike counterforce capability.

Pre-MAD Policy   The precise role of nuclear weapons in U.S. strategy was undefined in the years immediately after the bombing of Hiroshima and Nagasaki, but for U.S. leaders the primary objective was to maintain America's nuclear superiority. (Nuclear superiority is defined here as the ability to attack an adversary without fear of nuclear retaliation.) As General Leslie Groves, head of the Manhattan Project, wrote in 1946 as he looked to the future of a U.S.-Soviet arms competition, "the United States must for all time maintain absolute supremacy in atomic weapons."[33] General Dwight Eisenhower, Army Chief of Staff at the time, concurred, "We must be superior to any nation on any kind of weapon or equipment which we need."[34] President Harry S. Truman was initially inclined to seek international control of nuclear weapons through the Baruch Plan in the United Nations but had given up on this goal by 1949: "Since we can't obtain international control we must be strongest in atomic weapons," he told his advisers.[35] Although offense-defense theory, like all realist theory, would expect states to try to maintain unilateral military advantages, the degree to which American leaders sought to preserve nuclear superiority seems at odds with the view of nuclear weapons as the ultimate defensive technology.

Nuclear superiority was deemed so important in these years that the United States contemplated attacking the Soviet Union in order to maintain its monopoly. The history of America's flirtation with a preventive nuclear strike against the Soviet Union in the late 1940s and early 1950s is underappreciated by many today, but at the time the idea of preventive war was widely promoted by prominent military and political leaders, congressmen, journalists, and scholars.[36] The main rationale for preventive war was easy to grasp: if the brewing competition with the Soviet Union was likely to result in war, as many believed it inevitably would, it made more sense (and, for some, it was morally preferable) for the United States to fight while it was still in sole possession of the bomb.[37]

Preventive war logic came to the fore in 1948 as the prospect of a U.S.-Soviet confrontation increased. Even before the Berlin blockade, Winston Churchill was calling for a showdown with the Soviet Union before the American nuclear monopoly was broken, and he privately urged the United States to threaten nuclear attack if the Soviets did not withdraw from East Germany.[38] After the Soviets cut off access to Berlin, U.S. Defense Secretary James Forrestal and Air Force Chief of Staff Hoyt Vandenberg advised Truman to launch an immediate preventive attack on the Soviet Union. Truman resisted these calls but dispatched two squadrons of B-29s to Western Europe as an implicit atomic threat[39] and agreed to war plans that would entail a full-scale nuclear attack against the Soviet Union. Truman reserved for himself the final decision to use nuclear weapons but told his advisers that he would do so if the Soviets obstructed the Western airlift to Berlin.[40]

The Soviets did not provoke a war in 1948 because they understood that the U.S. nuclear monopoly ensured an American victory in any such war.[41] Nuclear superiority served as more than a deterrent to Soviet aggression, however; it also gave U.S. leaders greater freedom to maneuver. American policymakers understood that their nuclear monopoly allowed them to devote substantial economic resources to rebuild war-torn allies, avoid deploying massive numbers of conventional forces to Western Europe to defend against a Soviet ground invasion, and, most important, compel the Soviets to back down in any serious diplomatic crisis. As the historian Melvyn Leffler writes, "U.S. officials, therefore, possessed the confidence to do things they might otherwise have hesitated to do if they suspected that their actions could trigger a sequence of moves that might lead to war."[42]

The American nuclear monopoly came to an end when the Soviet Union tested its first atomic bomb in August 1949. Although it appears that the quest for nuclear advantage endured, one might argue that American policymakers no longer perceived any real importance in maintaining nuclear superiority. It is true that NSC-68, the famous grand strategy document prepared for President Truman in April 1950, which otherwise called for an aggressive response to the growing Soviet threat, explicitly ruled out the preventive war option.[43] Moreover, Truman himself claimed to reject preventive war on moral grounds.[44] And the simple fact remains that the United States did not launch a preventive war before or after it detected the Soviet atomic test. These arguments could be used to counter the claim that nuclear superiority mattered for American policymakers. Three points can be offered in response.

First, the preventive war option was officially set aside in 1950 primarily because of concerns about America's military weakness at the time, not because nuclear superiority was an unimportant objective or because preventive war was seen as morally reprehensible. The United States might have been able to destroy Soviet nuclear capabilities with a preventive attack in 1950, but most observers—including the main author of NSC-68, Paul Nitze—understood that the subsequent war in Europe would be long and costly (although all agreed it would eventually be won by the United States given its nuclear advantage). Thus, the mainstream position among military and civilian leaders was that the United States needed to build up its nuclear and conventional power before taking on a general war.[45] Although the idea of a preventive strike against the Soviet Union was vigorously discussed and debated in the preparation of NSC-68, it was eventually seen as the second-best option.[46] In the meantime, spurred by the start of the war in Korea in June 1950, the United States embarked on a massive rearmament program. Between 1950 and 1953, in addition to making the decision to build the hydrogen bomb, Truman tripled the rate of defense spending and nearly quadrupled the nuclear weapons stockpile.[47]

Second, preventive war thinking did not die with NSC-68. In one week in August 1950, for example, the secretary of the navy, Francis Matthews, and the commandant of the Air War College, General Orvil Anderson, publicly advocated preventive war against the Soviet Union.[48] Both officials were rebuked, but press reports speculated that Defense Secretary Louis Johnson, who advocated the preventive war in private conversations, had instigated Matthews's comments as a policy "trial balloon."[49] In November 1950, the Central Intelligence Agency (CIA) director told the National Security Council that the Chinese intervention had raised "the question as to what point the U.S. will be driven to, to attack the problem at its heart, namely Moscow, instead of handling it on the periphery as at present."[50] Several prominent senators, including Henry Jackson, John McClellan, and Paul Douglas, supported preventive war to varying degrees. The idea of preventive war grew more and more attractive as American military capabilities skyrocketed and the Korean War became protracted.[51] In January 1952, as the Korean War negotiations bogged down, Truman vented privately that unless the Communists settled for peace, "Moscow, St. Petersburg, Mukden, Vladivostok, Peking, Shanghai, Port Arthur, Dairen, Odessa, Stalingrad and every manufacturing plant in China and the Soviet Union will be eliminated."[52]

President Eisenhower entered office in 1953 with even fewer qualms than Truman about considering preventive war and resorting to atomic diplomacy. In May 1953, as the Korean armistice talks stalled, Eisenhower agreed to gradually escalate the war and approved contingency plans that included nuclear strikes against China, which he knew could lead to Soviet intervention and, in turn, the launch of America's all-out first-strike nuclear attack plan against the Soviet Union.[53] Although Communist concessions quickly brought the Korean War to an end, the American government's examination of the preventive war option continued. In September 1953, Eisenhower concluded that a nuclear arms race with the Soviet Union forced him "to consider whether or not our duty to future generations did not require us to *initiate* war at the most propitious moment that we could designate."[54] In May 1954, Eisenhower was briefed on a Joint Chiefs of Staff report proposing that the United States consider "deliberately precipitating war with the USSR in the near future," before Soviet nuclear capability became a "real menace."[55] In June 1954, Eisenhower again wondered aloud to his advisers whether "we should perhaps come back to the very grave question: Should the United States now get ready to fight the Soviet Union?"[56] Preventive war against the Soviet Union was discussed in other instances both before and after 1954,[57] but the option was rejected by American policymakers by the end of that year. No doubt moral considerations played a role in the decision not to attack preventively, but the most compelling rationale lies in a combination of U.S. strength and weakness at the time. On the one hand, American leaders perceived that their current and future strategic situation vis-à-vis the Soviet Union looked more promising than it had a few years earlier, even in the shadow of a mature Soviet arsenal. On the other hand, preventive war was also increasingly unattractive as policymakers could not avoid the conclusion that an attack on the Soviet nuclear arsenal would be followed by a long, costly war in Europe and staggeringly difficult occupation of Russia.[58]

American leaders also contemplated a preventive attack against China in the early 1960s. President John F. Kennedy wanted to halt China's emerging nuclear weapons program because he believed the Chinese "loom as our major antagonists of the late 60's and beyond" and "we would have a difficult time protecting the free areas of Asia if the Chinese get nuclear weapons."[59] Kennedy and his advisers thus pushed a massive intelligence effort to determine the full extent and nature of the Chinese nuclear program, initiated contingency plans for air attacks on Chinese nuclear facilities, and encouraged the CIA to explore covert and paramilitary options. We now know that Kennedy sought Soviet cooperation for joint U.S.-Soviet action on "limiting or preventing Chinese nuclear development" or, if the Soviets declined to participate, acquiescence of unilateral U.S. military action without the Soviets. These American overtures to the Soviets in 1963 and 1964 were rebuffed. Although planning and discussion about a preventive attack on China continued after 1964, the Soviet response, combined with Kennedy's assassination and President Lyndon B. Johnson's electoral concerns, largely put an end to the issue.[60]

Third, even as NSC-68 ruled out preventive war, the strategy document explicitly accepted the idea of *preemptive* attack: "The military advantages of landing the first blow . . . require us to be on the alert in order to strike with our full weight as soon as we are attacked, and, if possible, before the Soviet blow is actually delivered."[61] Throughout the 1950s, Eisenhower and the military establishment repeatedly emphasized the need to strike first upon warning of Soviet aggression. American nuclear policy became publicly known as "massive retaliation," but the actual strategy was one of massive preemption. Given warning of hostilities, the Strategic Air Command (SAC) war plan called for an immediate combined attack with large-yield thermonuclear weapons against all categories of targets—nuclear capabilities, military forces, and urban-industrial targets—within the entire Sino-Soviet bloc. The single massive blow would, according to an official summary of a SAC briefing, leave the Soviet Union "a smoking, radiating ruin at the end of two hours."[62] General Curtis Lemay, SAC commander, summed up the strategy when he explained, "If the U.S. is pushed in a corner far enough we would not hesitate to strike first."[63]

The Kennedy administration proclaimed a departure in nuclear policy from what it called a dangerously rigid "massive retaliation" strategy toward a less provocative "flexible response" doctrine. The basic idea, as Defense Secretary Robert McNamara explained, was to make "nuclear deterrence more credible" because previous threats of a full-scale nuclear strike against the Soviet Union would no longer be plausible or effective.[64] In practice little changed.[65] President Kennedy, like Truman and Eisenhower before, emphasized the importance of American nuclear superiority, maintained the capability and option to launch a full-scale preemptive nuclear attack, and capitalized on the fear of a U.S. nuclear first strike for coercive political purposes. In both the Berlin crisis of 1961–62 and the Cuban missile crisis of 1962, Kennedy signaled his intention to resort to nuclear arms to counter Soviet initiatives. In both cases the Soviets backed down in the face of military realities.

New evidence from the early 1960s validates Soviet fears of an American nuclear attack, as the United States repeatedly explored the first-strike option.[66] Whether the United States would ever have actually launched such a strike is less important to establish for present purposes than the fact that nuclear superiority gave the United States a politically meaningful advantage in the eyes of both American and Soviet leaders. The Berlin crisis is especially revealing. At the time, President Kennedy was deeply interested in the preemptive war option: he repeatedly discussed the issue with top political and military leaders; pressed hard for reports on the mechanics and effectiveness of a nuclear first strike, especially if the Berlin crisis were to escalate into armed conflict; and concluded that a preemptive strike was a viable policy option.[67] In June 1961, Kennedy told French president Charles de Gaulle, "In nuclear warfare, the advantage of striking first with nuclear weapons is so great that if the Soviets were to attack even without using such weapons, the U.S. could not afford to wait to use them."[68] According to the minutes of a meeting with the Joint Chiefs of Staff in June 1961, Kennedy "stated that the critical point is to be able to use nuclear weapons at a crucial moment before they [the Soviet Union] use them. He inquired as to our capabilities of making such a decision without letting the enemy know that we are about to do it."[69] This was more than just speculation. During the summer of 1961, as the Soviets threatened to seize West Berlin from the Western powers, the Kennedy administration drew up detailed plans for a nuclear first strike. A summary of the plan was forwarded to Kennedy in September, and he responded with a series of clarifying questions (e.g., "Assuming . . . the U.S. should launch an immediate nuclear strike against the Communist Bloc, does the JCS Emergency Actions File permit me to [do so] without first consulting with the Secretary of Defense and/or the Joint Chiefs of Staff?"; "What would I say to the Joint War Room to launch an immediate nuclear strike?").[70]

Kennedy and his advisers continued to seriously explore the first-strike option until September 1963. In the meantime, the Berlin crisis ended in large part because the Kennedy administration sent clear public and private warnings to the Soviets that the United States would not yield on the issue and that nuclear war was a real possibility if the crisis escalated. In particular, speeches by Kennedy (on July 25, 1961) and Deputy Secretary of Defense Roswell Gilpatric (on October 21) were designed to highlight the great strategic advantage held by the United States, something the Soviets could not deny.[71] The current conventional wisdom, even among historians previously skeptical of the extent of Kennedy's nuclear brinksmanship, is that Kennedy was indeed prepared to launch a nuclear war if push came to shove over Berlin.[72] Fortunately, both the Americans and the Soviets understood that U.S. nuclear superiority allowed the Americans to take a strong stand on Berlin, and thus the Soviets were forced to concede.

The Cuban missile crisis of October 1962 provides mixed evidence on whether nuclear weapons gave the United States a meaningful political advantage.[73] On the one hand, U.S. nuclear superiority was slipping away, and American leaders may well have recognized that their coercive bargaining leverage had been diminished.[74] On the other hand, it appears that Soviet perceptions of the nuclear balance gave the United States important coercive advantages. Consider three points that support the claim of the importance of nuclear superiority. First, Soviet Premier Nikita Khrushchev attempted to place missiles in Cuba in large part because he understood that U.S. nuclear superiority had forced him to back down over Berlin the previous year. Khrushchev thought that the missile deployment would improve the overall Soviet strategic position enough so that another inevitable showdown over Berlin would end differently. Second, U.S. leaders were willing to rely on threats to escalate to nuclear war for coercive purposes. At the height of the crisis, Kennedy publicly threatened a "full retaliatory response" if any of the missiles were launched from Cuba and stated emphatically that the United States would not tolerate the presence of missiles in Cuba. Third, it appears that the Soviet Union withdrew its missiles from Cuba rather than risk a nuclear war with the United States. Although the record shows Kennedy offered to remove U.S. missiles from Turkey and not invade Cuba, the Soviets made the decision to concede before news of this offer reached the Kremlin. In short, U.S. nuclear superiority loomed large over this crisis, as it had throughout the entire early phase of the cold war. In the words of two Soviet arms diplomats, "The Soviet leadership of the 1960s learned one thing too well: American superiority in the number of nuclear weapons led to an inequality that threatened Soviet security. . . . What the Soviet Union needed—vitally—was nuclear superiority to the American strategic arsenal. Indeed, had not the Cuban missile crisis presented itself as solid evidence of this fact?' "[75]

Policy in the MAD Era   On September 12, 1963, President Kennedy was presented with a report that led him to conclude that a preemptive strike on the Soviet Union was no longer a viable option. Even an all-out surprise first strike against the Soviet Union would leave enough surviving Soviet nuclear capability to cause about 30 million U.S. fatalities (in addition to at least 140 million Soviet fatalities). "The President asked whether then in fact we are in a period of nuclear stalemate," according to the summary of the meeting. "General Johnson [head of the committee writing the report] replied that we are. . . . The President again said that preemption was not possible for us."[76] Offense-defense theory would expect American policymakers to have responded to the fact that the United States no longer possessed a first-strike nuclear capability by accepting the status quo of a defense-dominant world and abandoning the pursuit of nuclear advantage. Instead, as a brief overview of U.S. nuclear policy after 1963 shows, the United States expanded its nuclear arsenal, improved its first-strike capabilities, and pushed to develop counterforce weapons and defensive systems.[77]

Offense-defense proponents find the American nuclear build-up puzzling or irrational because they believe that American leaders should have felt tremendously secure. Because the United States and the Soviet Union possessed secure retaliatory capabilities (i.e., MAD reigned) traditional military conquest was impossible, the security dilemma was virtually absent, and mutual security was readily available.[78] As a matter of actual policy, and not withstanding official public statements or rhetorical commitments to strategic stability in a MAD world, the United States built an enormous counterforce arsenal that threatened its adversary's retaliatory capability, took steps to limit damage to its own forces and society when practical, and generally persisted to seek ways to escape from MAD. Technological opportunism is better poised to explain this behavior. The United States appears not to have been content with the condition of mutual security entailed in MAD and felt great uncertainty about the pace and nature of technological change. The United States thus chose to compete for unilateral advantage and, quite possibly, sought absolute security through nuclear superiority.

Consider both changes in the size and nature of the U.S. nuclear arsenal during the MAD era. Recall offense-defense theory's prediction that nuclear adversaries should not be too concerned with comparing the relative size of their arsenals because even large shifts in relative force levels pose little threat to the "weaker" side's ability to retaliate and inflict unacceptable damage. The growth in size of the American nuclear arsenal starkly contrasts with this prediction, as U.S. forces far exceeded the capabilities necessary for assured retaliation. It is safe to assume that the United States lost its ability to completely disarm the Soviet Union with a massive first strike by 1963 at the very latest. By this date, the Soviets could deploy 638 strategic warheads on a triad of bombers, intercontinental ballistic missiles (ICBMs), and submarine-launched ballistic missiles (SLBMs).[79] More important, by this date American leaders believed that the United States could not effectively disarm the Soviets with a first strike. Even before President Kennedy expressed his doubts, President Eisenhower had (by his final years in office) largely abandoned hope of preventively disarming the Soviet Union.[80] The minimum U.S. force levels thought necessary to inflict unacceptable damage on the Soviet Union in a retaliatory strike were also well in hand by this time.[81] In 1964, the United States deployed 4,718 strategic nuclear warheads on its triad of bombers, ICBMs, and SLBMs, whereas the Soviets deployed about 800 strategic warheads.[82] Many analysts thought the United States would be able to inflict unacceptable damage on the Soviet Union with a far smaller arsenal.[83] Either way, according to offense-defense theory, security should have been plentiful for both superpowers. Yet the Soviet Union began a massive military buildup of both nuclear and

conventional forces, and the United States quickly followed with a vast increase of its own warheads. Eventually, the United States deployed 6,135
deliverable warheads in 1970; 10,768 in 1980; and 12,304 in 1990 at the end
of the cold war.[84]

The qualitative nature of the U.S. nuclear arsenal is as important as the
sheer numbers. According to offense-defense logic, once states find themselves in a MAD world, they should not attempt to gain an advantage at
the nuclear level by building offensive counterforce weapons—that is,
forces aimed at destroying an adversary's strategic nuclear weapons. In
the American case, policymakers declared a nuclear doctrine consistent
with MAD but fairly consistently embraced a counterforce posture and
strategy.[85] Paradoxically, U.S. counterforce planning and targeting began
in the 1950s, just when it was becoming apparent that the Soviet Union
would soon acquire a secure second-strike capability.[86] By the end of 1962,
the U.S. nuclear war plan had taken on an overwhelming counterforce
character, as 90 percent of the Soviet bloc targets were nuclear delivery
systems, antiair systems and bases, command-and-control centers, and nuclear weapons production and storage facilities.[87] The remainder of the
decade saw many U.S. analysts and policymakers becoming increasingly
convinced that damage limitation was a hopeless strategy.[88] Still, the
United States decided to deploy nationwide antiballistic missile (ABM)
defenses that might one day neutralize the Soviet retaliatory capability
and continued to enhance its counterforce arsenal by building increasingly accurate weapons capable of destroying hardened Soviet targets. In
the 1970s, despite having concluded agreements with the Soviet Union to
limit strategic defenses (more on that issue is discussed below), the
United States continued to enhance its counterforce arsenal by building
highly accurate weapons capable of destroying hardened Soviet targets.[89]
Among other steps, the United States built Minuteman ICBMs and Polaris
SLBMs, married highly threatening multiple independently targetable
reentry vehicles (MIRVs) to some of its ballistic missiles, and decided to
upgrade its Minuteman III ICBMs and deploy highly lethal and accurate
Peacekeeper MX ICBMs, Trident D-5 SLBMs, and Pershing II medium-
range ballistic missiles—all of which threatened the Soviet ability to retaliate in a nuclear exchange.[90] In the 1980s the Reagan administration took
counterforce to an extreme by pursuing effective strategic defenses and
offensive counterforce programs. Strategic defense efforts fell under the
Strategic Defense Initiative (SDI) and a host of air-defense, early warning,
and civil defense programs. Offensive programs included accelerating the
Trident D-5 program and building new bombers, cruise missiles, warheads, and sensors.[91]

It is worth noting that the Soviet Union also did not regard the possession of an assured destruction capability as sufficient to maintain its security. The Soviets began to build an ABM system around Moscow in the mid-1960s and, after signing the ABM Treaty in 1972, continued to develop its strategic defense capabilities through air defense programs against bombers and civil defense efforts aimed at protecting Soviet leadership. More important, the Soviets embarked on a massive strategic nuclear buildup that stressed offensive counterforce.[92] By 1986 about 80 percent of Soviet ICBMs were deployed on heavy, accurate MIRVed SS-18 and SS-19 missiles, which were specifically designed to destroy hardened military targets and the U.S. ICBM force.[93]

In taking these steps, the Soviets clearly understood and sought to counter the nuclear logic of our extended deterrence commitment to Western Europe. In order to protect American allies from a conventional Soviet attack the United States required a nuclear first-strike capability. Thus, the Soviet ABM and ICBM efforts were first and foremost aimed at neutralizing U.S. strategy and preserving the utility of the Soviet conventional advantage in Central Europe. Although the details of the strategic calculus differ somewhat from the U.S. case, the Soviet Union similarly built a robust counterforce arsenal during the cold war and sought to fight and "win" a nuclear war in the event such a war occurred.

Making Sense of Nuclear Arms Racing   At this point in the discussion two categories of important counterarguments should be addressed. First, offense-defense proponents might argue that America's pursuit of ever-larger numbers of strategic nuclear weapons with an increasingly heavy counterforce emphasis was simply irrational, a "wildly irrelevant technical competition" that poses a challenge for any rationalist theory (including technological opportunism).[94] Proponents might argue that America's irrational (or "suboptimal") behavior was driven by massive misperceptions about the (in)ability to retaliate in the event of a Soviet attack; about the level of capability needed to deter a Soviet attack; about the nature of the offense-defense balance of technology, which so heavily favored defenders that escaping from MAD was a pipe dream; and about the counterproductive result of U.S. policies, which mistakenly conveyed malign motives.[95] Second, proponents might argue that the United States eventually embraced the logic of mutual assured destruction in the late 1960s, 1970s, and early 1980s, with its agreement to ban missile defense systems and place limits on offensive forces. Both counterarguments have merit, and neither can be dealt with sufficiently here, but technological opportunism can offer a response.

Did U.S. leaders misperceive their own retaliatory and deterrent needs? If simply deterring a Soviet nuclear attack was the goal, then it does appear that American leaders were too pessimistic. Some hawkish advocates of an American buildup surely then miscalculated the size and nature of the U.S. nuclear force needed to ensure retaliation after a Soviet attack and exaggerated the degree to which U.S. forces needed to threaten Soviet interests to deter such an attack. (Of course, other policymakers might have advanced similar claims not because of misperceptions but for bureaucratic, organizational, or domestic political reasons.) However, although simple deterrence was a vital objective, there are good reasons to suspect that U.S. policy was driven by more ambitious goals, including the desire to attain nuclear superiority and remain ahead in an unpredictable technological race. One reason to think that the extensive list of offensive and defensive steps taken by the United States presented above was more in keeping with the goal of attaining nuclear superiority is that advocates of a larger and more threatening arsenal kept raising the bar on the stated criteria for an effective deterrent force as the cold war progressed. They ranged, for example, from ensuring that a percentage of Soviet industry and population would be destroyed (i.e., McNamara's original criteria) to countering the Soviet "throw-weight advantage" and civil defense programs to targeting Soviet command and control centers and leadership assets—all with the rhetorical justification that Soviet expansionist ambitions required the United States to hold hostage those assets Soviet leaders valued most.[96] America's embrace of MIRV technology in the 1960s and 1970s should also not be dismissed as driven by misperceptions of retaliatory or deterrent needs. To be sure, the strategic rationale for MIRV is enormously complex, but proponents advanced plausible arguments about the need to penetrate Soviet ABM systems, deter limited attacks against the U.S. homeland, extend deterrence to U.S. allies, and control escalation if war occurred.[97] In addition to the publicly stated rationales for MIRV, American planners thought seriously about how MIRV would contribute to a preemptive first-strike capability. According to one recently declassified official history of MIRVs, "From the time of its inception, MIRV was designed . . . to fulfill the U.S. requirements for enhanced penetrability and increased capability to attack a larger number of targets. *Although the avowed intent was the preservation of the U.S. deterrent capability, the issue of first strike capability was raised and widely discussed.*" The document explicitly notes that one of the advantages afforded by MIRV was "the enhancement of a first-strike capability."[98]

The counterargument about the irrationality of trying to escape from MAD given the tremendous defense-dominance of nuclear technology is powerful but again fails to appreciate the enormous uncertainty and strong incentives to compete faced by states when it comes to issues of technological change and national security. Technological opportunism explains why U.S. leaders were acting within the bounds of rational behavior when they pushed hard to develop the technology and capability to escape MAD. This opportunistic drive for nuclear advantages, however, did slacken during the period of strategic parity in the 1970s, even if it never went away.[99] While that evidence is also consistent with offense-defense theory, alternative explanations—for example, that the United States shifted its defense emphasis to conventional forces—abound. In a world of tremendous uncertainty about the motives of adversaries and the future course of technological progress, few states are willing to be content with the status quo of a defense-dominant world, especially when the prospect of total annihilation is balanced against the glimmer of absolute superiority.

A final counterargument stemming from the role of misperceptions contends that the United States failed to see that its actions mistakenly conveyed malign motives. This argument takes for granted that American objectives were defensive, status–quo-oriented, and benign—whatever those terms might mean—which is an assumption offense-defense theory makes about most states.[100] If the only thing the United States needed to do in the cold war was deter an attack on the U.S. homeland, then it might have happily settled for a level of minimal deterrence often promoted by offense-defense proponents. However, offense-defense theorists seem to overlook the fact that the scope of American interests since 1945 has always embraced more than this minimal objective of homeland defense. For one thing, the problem of extended deterrence—specifically, of deterring the Soviet Union from launching a conventional invasion of Western Europe—loomed large throughout this period and greatly contributed to America's pursuit of counterforce weapons and targeting. The Soviets' own assured destruction capability reduced the credibility of U.S. threats to use nuclear weapons in response to Soviet aggression, which undermined the U.S. ability to extend deterrence to its allies. Simply put, the logic of extended deterrence required a first-strike capability.[101] The problem for offense-defense theory is larger than the issue of extended deterrence, however. Most fundamentally, as technological opportunism (and broader offensive realism) predicts, the United States was not satisfied with the capabilities necessary to protect the status quo. It sought to convey a willingness to risk nuclear war to pursue a set of global interests, including the rolling back of Soviet power and influence around the world. Arguably, these goals included the desire to attain nuclear superiority and decisive power advantage over any great power rival. In short, the U.S. pursuit of nuclear superiority was part and parcel of America's global hegemonic ambitions.

Ultimately, the counterarguments about the irrationality of U.S. nuclear policy may not appreciate the extent to which the mere pursuit of technological superiority itself brought concrete military and political advantages to the United States. At a minimum, for example, U.S. nuclear policies gave American leaders a degree of influence over the ongoing defense policies of the Soviet Union, as the Soviets felt compelled to compete with the United States at almost every level of the nuclear competition. (Note, however, that the Soviets may have had a similar influence over U.S. defense policies, as the Soviet deployment of SS-18s created significant concerns among U.S. planners and energized schemes to catch up with Soviet "throw-weight" levels and prompt "hard-target-kill" capabilities.) At a maximum, America's refusal to accept the constraints and opportunities of a MAD world helped it win the cold war, as the arms competition eventually bankrupted the Soviet Union.

Aside from counterarguments about misperception, which (if correct) would undermine the apparent explanatory advantage of technological opportunism in the nuclear case, proponents of offense-defense theory might seek more direct support for their theory by pointing to the fact that the United States agreed to pursue limits on ABM systems and, in 1972, signed the ABM treaty with the Soviet Union. The U.S. decision to limit ABM systems certainly suggests that American leaders may have finally embraced mutual vulnerability as the best way to prevent war and avoid a dangerous and costly arms race. That behavior would be consistent with the logic of offense-defense theory, not technological opportunism. From the perspective of technological opportunism, however, the ABM treaty makes sense if American leaders had come to believe that the technology to effectively defend against Soviet missiles was not yet available. Thus, if the United States could not yet deploy effective defenses, it was in America's interest to prevent the Soviet Union from doing so, perhaps to leave the door open for U.S. first-strike planning.[102] This argument might be dismissed if not for the fact that U.S. policies before and after the negotiation of the ABM treaty displayed a desire to escape the constraints of a MAD world. Despite rhetoric about purely defensive and deterrent motivations behind earlier efforts to develop and deploy ABM defenses, American leaders wanted strategic defenses in order to limit the damage that would occur should the United States and the Soviet Union fight a nuclear war.[103] As Thomas Schelling notes, "The defenses that the United States and the Soviet Union were trying to develop in the 1960s and early 1970s were not really defensive in orientation. They were complements to an offensive force."[104] Of course, whatever one makes of the offensive or defensive motivations behind Reagan's decision to pursue SDI, the president's explicit and public rejection of mutual assured destruction cannot be squared with the predictions of offense-defense theory.

Conclusion

The invention of nuclear weapons in the second half of the twentieth century exponentially increased the level of firepower available to states. Nuclear weapons differ from all previous military technologies in that they offer the potential to inflict massively lethal damage on civilian populations and armies at the immediate outset of a military conflict. The firepower revolution of the late nineteenth and early twentieth centuries was mostly manifested at the tactical level of warfare, but the nuclear revolution is essentially a strategic phenomenon—that is, the scale and speed of destruction inherent in nuclear weapons renders tactical or operational issues related to the waging of battles and campaigns irrelevant. Consequently, the nuclear revolution is highly germane to offense-defense theory, which seeks to understand technology's impact on ultimate military outcomes, perceptions of these strategic effects, and the security behavior that follows from these perceptions.

Offense-defense theory holds that the emergence of nuclear weapons decisively shifted the balance of technology in favor of defense. Proponents of the theory would not expect nuclear war to be long and indecisive, as with other defensive shifts in the balance, but they would expect it to bring so much destruction so rapidly that no state would ever initiate it in the first place. In a world of mutual assured destruction capabilities, nuclear weapons are the ultimate defensive technology because they virtually eliminate the security dilemma and render war obsolete. The logic of this claim is essentially sound, although it muddies the conceptual waters of offense-defense theory by shifting the basis for coding technology's impact on the balance in strictly military operational terms. Nevertheless, and despite the fact that nuclear weapons give any sufficiently armed attacker the ability to destroy an adversary, it would seem that military victory (defined in terms of successful conquest) is impossible among nuclear states because an equally well-armed defender can destroy the attacker in turn. According to offense-defense theory, defenders can feel secure because deterrence through mutual assured destruction is easy to achieve and easy to maintain no matter what the adversary does to improve its first-strike capability. Three concrete and testable predictions about political outcomes flow from nuclear defense dominance. First, wars among nuclear powers should not occur. Second, security competition in regions of tertiary importance should not be intense. Third, arms racing beyond MAD requirements should not occur.

Overall, the empirical evidence from the cold war cuts against offense-defense theory. The United States and the Soviet Union quickly perceived their mutual vulnerability to nuclear destruction. Political and military leaders in each country correctly perceived the basic features of nuclear defense dominance. Indeed, the predicted absence of war between the nuclear superpowers fares best. However, other explanations can at least partly account for the "long peace." More important, the role of nuclear weapons in keeping the peace is undermined by several cases where nuclear weapons did little to restrain aggressive behavior. States armed with nuclear weapons have been attacked (Israel, by Egypt and Syria in 1973), states have intervened in wars involving nuclear powers (China in Korea against the United States in 1950), states possessing nuclear weapons have clashed militarily (China and the United States in 1969) and, arguably, nuclear-armed states have fought a war or something approximating one (India and Pakistan in 1999).

The evidence most clearly contradicts offense-defense predictions about competition in the third world and arms racing. The United States and the Soviet Union should have felt tremendously secure behind their nuclear arsenals. Nevertheless, the superpowers competed hard in regions of little importance to their core security interests and engaged in an intense and costly arms race. The numbers of weapons accumulated in the superpower arsenals far surpassed any reasonable estimate of the levels needed to ensure a second-strike capability, and both sides pursued large offensive counterforce programs. In sum, the security policies of the United States and the Soviet Union during the cold war pose a significant challenge to offense-defense theory.

Conclusion

Technology and the Primacy of Politics

Offense-defense theory is an optimistic theory because it holds that many wars are driven by avoidable causes. The theory contends that international competition, conflict, and war are less likely to occur when the prevailing nature of technology favors defenders over attackers. When defense dominates, states can feel secure and act benignly. Preemptive wars to strike an adversary before being struck become unnecessary. Preventive wars to eliminate potential rivals or deal with grave and gathering dangers become less compelling. Even wars of pure aggression, expansion, and conquest become more difficult when defense is easier than offense. To be sure, highly aggressive states or states with greater power than their adversaries will always be tempted to fight wars in the face of the strength of the defense or seek ways to get around the constraints of defense dominance. But offense-defense theory points the way for cooperation among actors who are relatively satisfied with the status quo and mainly concerned with preserving their power and protecting their security. Moreover, according to offense-defense proponents, because actual offense dominance is a rare occurrence in history, most wars are avoidable if states can correctly perceive that the objective offense-defense balance favors defenders.

If offense-defense theory is correct, however, there are major reasons to be concerned about contemporary international politics. First, the increasing threat of nuclear terrorism by radical Islamic groups undermines the relative optimism of offense-defense theory's understanding of the nuclear revolution. The cold war-era idea that robust deterrence among nuclear states serves as the functional equivalent of defense dominance would not seem to apply in situations where states are interacting with holy warriors bent on martyrdom. If this is the case, the next sixty years of the nuclear era are not likely to be as peaceful. Second, if one can translate the logic of conventional military operations to the realm of terrorist and counterterrorist operations, it may be that offense now has the advantage. American military might prevents terrorists from fighting the United States on its own terms. Thus, as Richard Betts writes, "To smite the only superpower requires unconventional modes of force and tactics that make the combat cost exchange ratio favorable to the attacker."[1] In other words, surprise terrorist attacks offer greater bang for the buck than passive defenses. The implication of this is that the United States and its allies should emphasize counteroffensive operations—including preemptive and preventive attacks—rather than rely on homeland and other passive defenses to defeat terrorists. The downside, of course, comes when the benefits of counteroffensive operations against terrorists come at the price of alienating civilian populations and sapping political support through the brutal tactics and collateral damage that often accompany these operations.[2]

The arguments presented in this book undermine offense-defense theory, although the evidence and alternative explanations advanced are unlikely to offer any additional comfort. Despite its wide use in international relations scholarship and potential relevance for arms control and other related policies, the offense-defense balance of military technology is not a powerful causal variable. Explanations built on the concept of a balance provide little leverage for understanding either military outcomes or political decision making in international politics. Below I summarize the general and specific findings of my research, discuss the implications of my critique, and point to areas deserving further theoretical development and empirical testing.

The Offense-Defense Balance and Military Outcomes

Is there an offense-defense balance of technology that can be used to explain or predict *military outcomes*? In answering this question one can readily acknowledge that the nature of warfare at any given time reflects the pool of technology and weapons available to adversaries involved in conflict. Railroads permitted states to deploy and sustain far larger armies than ever before. Machine guns inflicted unprecedented losses on assaulting infantry. Tanks came to dominate the battlefield like no other weapons system. Nuclear weapons are so destructive that their use in war would likely render battlefield victories meaningless. Yet it is difficult to categorize the impact of technological change in offense-defense terms. Even when shifts in the offense-defense balance of technology are discernible, their effects on combat outcomes are contingent and usually not profound.

There are at least two major reasons why the offense-defense balance of technology is typically ambiguous and insignificant as a determinant of military outcomes. First, the variance between offensive and defensive advantages is often trivial. Military experts concur with Clausewitz that offense is intrinsically harder than defense because offensive operations usually require more forces and higher costs than defensive operations.[3] The offense-defense balance thus must take an extreme value—a very large offensive or defensive advantage—to shift this baseline condition of military operations and have a noticeable effect on battlefield outcomes. However, such large shifts in the balance are quite rare in history. The record shows not only that military technological innovation has been more evolutionary than revolutionary but also that decisive technological advances have spread quickly among states.[4] This means that asymmetric technological superiority is typically fleeting and, more important for offense-defense theory, offensive or defensive advantages do not endure long before antidotes emerge.[5] For example, tanks emerged simultaneously with tools of containing or defeating tanks, including specially designed antitank guns or other tanks.[6] If, as one proponent of offense-defense theory writes, the "offense-defense balance is most likely to be significant . . . when it changes dramatically or when it confers a very large advantage on the offense or defense," then the evidence presented in the previous chapters demonstrates that the balance is rarely significant.[7]

The second reason why it is difficult to characterize technology's impact on warfare in offense-defense terms is that the outcomes of military conflicts are far more likely to be determined by other important variables. Specifically, the distribution of power (the quantitative and qualitative balance of military forces) and the difference in military skill (in terms of doctrine, strategy, operations, and tactics) almost always have a far greater influence on the course of war and the results of conflict than the offense-defense balance. Proponents of offense-defense theory argue that all three variables (power, skill, and the offense-defense balance) are complementary, that each variable has the potential to overwhelm the others under certain conditions, and that the offense-defense variable has been neglected or misunderstood.[8] By contrast, the evidence presented here suggests that the offense-defense balance has not been underappreciated because it rarely exerts a powerful and independent effect on the course of battle and results of war. In short, the offense-defense balance adds little to explanations of military outcomes when compared with the variables of relative power and skill.

The combination of these two factors—that the offense-defense balance does not vary much and that other factors wash out the effects of the balance—also explains why measuring the balance of technology in practice is difficult, if not impossible. In this book I derived the most useful, clearly articulated, logically sound, and frequently employed criteria for how changes in military technology shape the offense-defense balance. I then applied these criteria in an analysis of four key historical cases. Even the largest shifts in the values of these criteria did not result in clear, profound, and predicted shifts in the relative ease of attack and defense in war. For example, possible shifts at one level of warfare (say, tactical) often did not translate into similar movements in the balance at a higher level (e.g., strategic or operational). Moreover, the most significant weapons and military systems tend to combine numerous technologies. These bundles of technologies—for example, the tank's combination of mobility, firepower, and protection—enormously complicate the task of assessing the offense-defense balance. Again, even when large shifts in the offense-defense balance are detectable—as with the nuclear revolution (in theory) or with the small arms and artillery revolution of the late nineteenth and early twentieth centuries—scholars have been unable to measure the balance or resolve important conceptual and empirical anomalies in these cases. Proponents of offense-defense theory contend that a cost-ratio measurement of the offense-defense balance can be calculated using detailed net-assessment techniques, but no such procedure has yet been conducted. In sum, this study concludes that an offense-defense balance of technology cannot be used in any meaningful way to explain or predict military outcomes.

The Offense-Defense Balance and Political Outcomes

Do perceptions of the offense-defense balance affect *political decisions* to initiate conflict? This question gets to the heart of whether offense-defense theory is, or can be, a fruitful theory of international politics. Even if one accepts the argument that an objective offense-defense balance is extremely difficult to measure and ultimately insignificant as a determinant of military outcomes, *beliefs* about such a balance could still act as an important influence on political decision making and military planning. However, the historical evidence shows that when it comes to understanding the military implications of new technologies, military and political leaders generally "get it right." That is, although they do not think in strict offense-defense terms, leaders do not typically suffer from major misperceptions about what they have experienced with new military innovations in training, experiments, or battle. The obvious example is perceptions of the impact of firepower on tactics before World War I. German military experts, as well as other European observers, understood correctly that the small arms and artillery revolution had made infantry assaults against prepared defenders enormously more costly than in the past. The historical record also shows that perceptions of the balance of technology do not play a significant role in deterring or encouraging political decisions to initiate war. Germany, to continue the example, did not find greater security in the perceived defensive advantages of modern firepower. Instead, it sought a strategy whereby it could defeat its enemies in a quick and decisive fashion and avoid a prolonged war of attrition. That it never developed such a strategy—that German leaders sent their forces into battle in 1914 without any real expectation that the coming war would be anything but long and protracted—is all the more reason to doubt the political consequences of the offense-defense balance.

What accounts for this disconnect between perceptions of the nature of technology and political decision making? Scholars have demonstrated that military doctrines and strategies are often more responsive to military organizational biases and statesmen's perceptions of international politics than to the implications of prevailing weapons technologies.[9] I would take this a step further: political and military elites tend not to shape their strategies on the basis of military technology but rather to view the utility of new technologies through the lens of their current strategies. In short, leaders want to know how given technologies can aid their strategy, not how their strategy should change in light of these technologies. In the cases examined in this book, perceptions of the international distribution of material power played a large role in explaining offensive strategies and war initiation, whereas perceptions of the nature of technology were relatively unimportant. German leaders acted aggressively in 1864, 1866, 1870, 1914, and 1939–41 because they wanted to expand Germany's relative power in Europe before the growing power of other states would threaten Germany,[10] not because they believed that technological offensive advantages had made those rivals more threatening. In sum, perceptions of an offense-defense balance of technology have little effect on political decision making in security affairs.

Theory and Evidence

Offense-defense theory contends that the relative ease of attack and defense often plays a major role in causing instability and war in international politics. The theory holds that states will tend to seek security through aggression when offensive advantages render the military capabilities of others more threatening, whereas peace is more likely when defensive advantages make changes in the balance of military power less worrisome.

An empirical evaluation of offense-defense theory predictions must differentiate between the "broad" and "core" version of the theory. The lion's share of past criticism has been directed at the conceptual and operational problems endemic to broad versions of offense-defense theory—those versions that define the offense-defense balance to include a list of factors in addition to military technology. Although proponents believe the expanded definition improves the theory, in practice the broad version of the balance appears as an ad hoc collection of variables employed pell-mell to account for empirical anomalies and support theoretical qualifications. I argue that the core version of offense-defense theory, which focuses almost exclusively on how technology shapes the relative ease of attack and defense, is the potentially more fruitful approach. The core theory offers two basic criteria for judging how a given technology affects the offense-defense balance and thus military outcomes: mobility-improving innovations generally favor offense and result in quicker and more decisive victories for the attacker, whereas firepower-enhancing innovations typically strengthen defense and lead to more indecisive warfare. In terms of political outcomes, the theory predicts that states are more likely to initiate conflict when they perceive that the offense-defense balance favors offense.

The evidence can be summarized as follows. In terms of *military outcomes,* the mobility-favors-offense prediction was not supported in either of the two relevant cases. Though the greater strategic mobility of railroads allowed attackers to quickly concentrate larger forces at the frontier, better operational mobility helped defenders rush forces to threatened points and retreat in good order. Tanks gave attackers and defenders greater operational and tactical mobility but did not give offense or defense a clear advantage on the battlefield. The firepower-favors-defense prediction found some support in both cases but with major qualifications. Great advances in small arms and artillery gave defenders an advantage, but this occurred primarily at the tactical level of warfare and was dependent on doctrinal choice. Nuclear weapons, if used in war, would prevent meaningful territorial conquest. In terms of *political outcomes,* offense-defense predictions were not supported in any of the cases. Political and military leaders generally understood correctly the impact of technological innovations on the nature of warfare, but these perceptions did little to either ameliorate or exacerbate state security concerns.

Technological Opportunism and the Primacy of Politics

Given the pervasive influence of offense-defense theory in international security scholarship, the logical consistency and empirical validity of the theory might deserve further evaluation. The evidence discussed here, however, clearly suggests that scholars have overstated both the degree to which the nature of technology shapes military outcomes and the influence that beliefs of offense or defense dominance have on political and strategic decisions. What are the policy and theoretical implications of these findings?

Offense-defense theory claims prescriptive utility because misperceptions and miscalculations of the offense-defense balance often lead states to initiate conflict when they otherwise might feel secure with the status quo; these misperceptions can sometimes be ameliorated through arms control and confidence-building measures. By contrast, this book suggests that arms control or other measures aimed at correcting misperceptions of the offense-defense balance will generally be misguided. In no case examined in this book—nor in any other case, to my knowledge—was a state with "benign" intentions forced to adopt aggressive policies or start a war because of the perceived or misperceived offense-dominance of technology. Moreover, although some offense-defense proponents argue that states can shift the offense-defense balance toward defense dominance through national foreign and military policies, the sounder offense-defense explanations acknowledge that the objective balance of technology—as an external constraint on state behavior—is not directly malleable through state policies.

That is not to say that all arms control efforts are fruitless or that other policies aimed at preventing war are hopeless. To the contrary, challenging the primacy of technology as a cause of war implies that we should refocus attention on the real political issues at stake in international conflict. Arms control agreements may not make war less likely, but they might allow states to spend more of their resources on other pressing international or domestic problems, many of which can exacerbate international relations. International diplomacy, bargaining, and conflict resolution are more likely to succeed when these efforts look beyond technological fixes to the clashing political and strategic interests that increase the probability of war.

The arguments of this book also have important implications for the debate between offensive and defensive realist approaches to understanding international politics. As discussed in the introduction, the offense-defense balance is the key variable allowing defensive realists to argue that the strong propensity of states to compete is not an inevitable consequence of the structure of the international system. For defensive realists, much international conflict is driven by the operation of the "security dilemma"—the way in which an attempt by one state to increase its security has the unintended effect of decreasing the security of others and, through a spiral of action-reaction, undermining its own security. Some states, to be sure, will harbor aggressive designs on other states. In these cases, when differences are irreconcilable, defensive realists see conflict as unavoidable. In many other cases, however, defensive realists contend that if power is distributed roughly equally and the balance of technology does not heavily favor offense, states may feel secure and develop military postures that signal their peaceful intentions, thereby minimizing the outbreak of unnecessary conflict. In short, these scholars argue, the nature of the offense-defense balance determines the severity of the security dilemma and the level of conflict and cooperation in international politics.

Offensive realists reject the security dilemma as a model of international politics. In their view, the nature of the international system fosters competition, conflict, aggression, and war. These unfortunate consequences do not occur because the nature of technology forestalls the pursuit of mutual security but instead because state security requirements are frequently incompatible and opportunistic aggression is often a good avenue to acquiring greater power. The outbreak of war can better be explained by the shifting distribution of power among states than changes in the offense-defense balance.

My findings buttress offensive realism. The goal of this book was not to directly test offensive realism against defensive realism, but my analysis shows that defensive realist explanations rest on a shaky foundation. If, as shown in chapters 2–5, the offense-defense balance of technology is a flawed concept—one that has difficulty explaining military or political outcomes—then the security dilemma model is also flawed. If changes in the balance of technology are rarely perceptible or insignificant, then the security dilemma seldom varies, and explanations built on it cannot capture fluctuations in the occurrence of war and peace. Above all, the evidence presented here shows that perceptions of technological constraints on offensive operations do little to diminish a state's eagerness and willingness to gain power at the expense of other states when the opportunity arises.

This conclusion points to a promising avenue for future research. Is there an alternative explanation for the relationship between technological change and international politics? In the introduction I referred to "technological opportunism"—the explanation that states rarely view new technological developments as a means to preserve the status quo or signal benign intentions but instead seize on even purportedly defensive technologies as potential opportunities to gain military and political advantage over rivals. Technological opportunism attempts to make explicit arguments about technology that previous offensive realist explanations leave implicit. Such an explanation might help account for several patterns in international politics: the lack of evidence that states attempt to signal their benign intentions through the types of forces they produce and deploy; the infrequency with which states ignore opportunities to gain relative power over their rivals; the frequency with which states develop technologies and weapons even when they are likely to be perceived as particularly threatening to others and thus destabilizing. In short, technological opportunism may help us understand why states tend not to be dissuaded from bolstering their capabilities by fear of making other states insecure.

Technological opportunism appears to be far more consistent with the empirical evidence than offense-defense theory. Consider the historical cases examined in this book. Helmuth von Moltke, Otto von Bismarck, and other top military and political leaders in Prussia believed that the spread of railroads would continue to strengthen the defender—yet they increasingly pursued aggressive political goals and provoked three wars in less than a decade. Alfred von Schlieffen and, especially, the younger Helmuth von Moltke, as well as their fellow officers and military observers at the time, fully understood that the vast improvement in firepower technology embodied in small arms and artillery had placed attacking forces at a great disadvantage to defenders—yet they sought to overcome these obstacles by trying to develop a plan to quickly defeat Germany's neighbors. Contrary to blitzkrieg lore, Adolf Hitler and his generals possessed no new radical doctrine of armored warfare in 1940, nor did they necessarily believe that tanks had shifted the offense-defense balance in favor of attackers—yet Hitler launched World War II by attacking Poland and decided to attack France even before the emergence of a bolder plan based on armored mobility and strategic surprise. Finally, military and political leaders in the United States and the Soviet Union understood that the nuclear revolution had created conditions of mutual assured destruction and enormous defense dominance—yet they relentlessly built larger and more threatening arsenals and competed hard in regions with little strategic value. All of this evidence undermines offense-defense theory, whereas technological opportunism appears to offer a more promising framework for understanding the same behavior.

Prior to the development of offense-defense theory in the field of international politics, Bernard Brodie wrote, "There seems not to be any direct proportionality between technological change and military-political consequences, even though we acknowledge that historically there has been a close relationship between one and the other."[11] Over a quarter century later, Brodie's point still stands. Understanding the limitations of offense-defense theory might help scholars develop more nuanced explanations for, and construct more precise empirical tests on, the role of technology in fostering international stability and conflict. Ultimately, however, the theory may simply not provide enough analytical leverage to be useful for understanding international politics, especially given the complexity of measuring the offense-defense balance of technology.

Contemporary scholars and policymakers face numerous security issues that revolve around the military and political implications of new technological innovations. Will a so-called revolution in military affairs (RMA) allow the United States to maintain its global hegemonic position at low cost for a long time to come? Will the spread of weapons of mass destruction and ballistic missile technology beyond the great powers create instability or foster peace? Will a U.S. missile defense system upset the nuclear balance, set off a global arms race, or encourage the threats such a system is designed to eliminate? What are the implications of technological asymmetry for the current war on terrorism? Given these pressing issues and debates, it should be clear that the relationship between technological change and international security is too important and fascinating a subject to abandon just because the dominant theory of that relationship is deeply flawed.

It would also be a mistake to reject a competing theory simply because of its pessimism about the future of international relations. Despite recurrent claims that world politics have been (or are being) transformed, security competition and war are enduring features of international life and technological change will continue to be seen by states as a potential means of gaining power over each other. Acknowledging this gloomy reality, as well as seeking ways to stem its negative consequences, is difficult. But the goal of making the world a better place is no better served by the specious promise of a technological peace.

Bibliography

Abdil, George B. *Civil War Railroads*. Seattle: Superior Publishing Co., 1961.

Achen, Christopher H. and Duncan Snidal. "Rational Deterrence Theory and Comparative Case Studies." *World Politics* 41 (January 1989): 160–61

Adams, Karen Ruth. "Attack and Conquer? International Anarchy and the Offense-Defense-Deterrence Balance." *International Security* 28 (Winter 2003–4): 45–83.

Addington, Larry H. *The Blitzkrieg Era and the German General Staff, 1865–1941*. New Brunswick, N.J.: Rutgers University Press, 1971.

——. *The Patterns of War since the Eighteenth Century*. Bloomington: Indiana University Press, 1994.

Adriance, Thomas J. *The Last Gaiter Button*. New York: Greenwood Press, 1987.

Albertini, Luigi. *The Origins of the War of 1914*. 3 vols. London: Oxford University Press, 1952.

Andreski, Stanislav. *Military Organization and Society*. London: Routledge and Keegan Paul, 1968.

Angevine, Robert G. *The Railroad and the State: War, Politics, and Technology in Nineteenth-Century America*. Stanford: Stanford University Press, 2004.

Ball, Desmond. "The Development of the SIOP, 1960–1983." In *Strategic Nuclear Targeting*, edited by Desmond Ball and Jeffrey Richelson, 57–83. Ithaca: Cornell University Press, 1986.

Ball, Desmond, and Robert C. Toth. "Revising the SIOP: Taking War-Fighting to Dangerous Extremes." *International Security* 14 (Spring 1990): 65–92.

Baumgart, Winfried. *The Crimean War, 1853–1856*. New York: Oxford University Press, 1999.

Bean, Richard. "War and the Birth of the Nation State." *Journal of Economic History* 33 (March 1973): 207–21.

Beckman, Peter R., Paul W. Crumlish, Michael N. Dobkowski, and Steven P. Lee. *The Nuclear Predicament: Nuclear Weapons in the Twenty-First Century.* 3rd ed. Upper Saddle River, N.J.: Prentice Hall, 2000.

Bell, P. M. H. *The Origins of the Second World War in Europe.* London: Longman, 1986.

Bellamy, Christopher. *The Evolution of Modern Land Warfare: Theory and Practice.* London: Routledge, 1990.

Bennett, Andrew, and Alexander L. George. "Case Studies and Process Tracing in History and Political Science: Similar Strokes for Different Foci." In *Bridges and Boundaries: Historians, Political Scientists, and the Study of International Relations,* edited by Colin Elman and Miriam Fendius Elman, 137–66. Cambridge: MIT Press, 2001.

Bennett, Scott D., and Allan C. Stam III. "The Duration of Interstate Wars, 1816–1985." *American Political Science Review* 90 (1996): 239–57.

Berman, Robert P., and John C. Baker. *Soviet Strategic Forces: Requirements and Responses.* Washington, D.C.: Brookings Institution, 1982.

Betts, Richard K. "Must War Find a Way? A Review Essay." *International Security* 24 (Fall 1999): 166–98.

——. "The Soft Underbelly of American Primacy: Tactical Advantages of Terror." *Political Science Quarterly* 117 (Spring 2002): 19–36.

——. *Surprise Attack: Lessons for Defense Planning.* Washington, D.C.: Brookings Institution, 1982.

Biddle, Stephen Duane. "The Determinants of Offensiveness and Defensiveness in Conventional Land Warfare." Ph.D. diss., Harvard University, 1992.

——. *Military Power: Explaining Victory and Defeat in Modern Battle.* Princeton: Princeton University Press, 2004.

——. "Rebuilding the Foundations of Offense-Defense Theory." *Journal of Politics* 63 (August 2001): 741–74.

——. "Recasting the Foundations of Offense-Defense Theory." Paper prepared for delivery at the annual meeting of the American Political Science Association, Boston, 1998.

——. "Testing Offense-Defense Theory: The Second Battle of the Somme, March 21 to April 9, 1918." Paper presented at the annual meeting of the American Political Science Association, Atlanta, 1999.

Biddle, Tami Davis. "Handling the Soviet Threat: 'Project Control' and the Debate on American Strategy in the Early Cold War Years." *Journal of Strategic Studies* 12 (September 1989): 273–302.

Bidwell, Shelford, and Dominick Graham. *Fire-Power: British Army Weapons and Theories of War, 1904–1945.* London: George Allen & Unwin, 1982.

Bijker, Wiebe E., Thomas P. Hughes, and Trevor J. Pinch, eds. *The Social Construction of Technological Systems: New Directions in the Sociology and History of Technology.* Cambridge: MIT Press, 1999.

Bishop, Denis, and W. J. K. Davies. *Railways and War since 1917.* London: Blandford, 1974.

Bismarck, Otto von. "Otto von Bismarck: Letter to Minister von Manteuffel,

1856." *Modern History Sourcebook: Documents of German Unification, 1848–1871.* http://www.fordham.edu/halsall/mod/germanunification.html.

Black, Robert C. III. *The Railroads of the Confederacy.* Chapel Hill: University of North Carolina Press, 1998.

Blacker, Coit. "The Kremlin and Detente: Soviet Conceptions, Hopes, and Expectations." In *Managing U.S.-Soviet Rivalry,* edited by Alexander George, 119–38. Boulder, Colo.: Westview, 1983.

Blaker, James R. *Understanding the Revolution in Military Affairs: A Guide to America's 21st Century Defense.* Washington, D.C.: Progressive Policy Institute, 1997.

Blechman, Barry M. *Technology and the Limitation of International Conflict.* Lanham, Md.: University Press of America, 1989.

Blechman, Barry M., and Robert Powell. "What in the Name of God Is Strategic Superiority?" *Political Science Quarterly* 97 (Winter 1982–83): 589–602.

Boemeke, Manfred F., Roger Chickering, and Stig Förster, eds. *Anticipating Total War: The German and American Experiences, 1871–1914.* Washington, D.C.: German Historical Institute, 1999.

Boggs, Marion William. *Attempts to Define and Limit "Aggressive" Armament in Diplomacy and Strategy.* Vol. 16, no. 1, *The University of Missouri Studies.* Columbia: University of Missouri, 1941.

Bond, Brian. *British Military Policy between the Two World Wars.* New York: Oxford University Press, 1980.

——. *Liddell Hart: A Study of His Military Thought.* New Brunswick, N.J.: Rutgers University Press, 1977.

——. *War and Society in Europe, 1870–1970.* New York: St. Martin's Press, 1983.

Bond, Brian, and Martin Alexander. "Liddell Hart and De Gaulle: The Doctrines of Limited Liability and Mobile Defense." In *Makers of Modern Strategy: From Machiavelli to the Nuclear Age,* edited by Peter Paret, 598–623. Princeton, N.J.: Princeton University Press, 1986.

Boserup, Anders. "Mutual Defensive Superiority and the Problem of Mobility along an Extended Front." In *The Foundations of Defensive Defence,* edited by Anders Boserup and Robert Nield, 63–78. New York: St. Martin's Press, 1990.

Boserup, Anders, and Robert Neild, eds. *The Foundations of Defensive Defence.* New York: St. Martin's Press, 1990.

Boutwell, Jeffrey, and Michael T. Klare. *Light Weapons and Civil Conflict: Controlling the Tools of Violence.* Lanham, Md.: Rowman & Littlefield, 1999.

Bracken, Paul. "America's Maginot Line." *Atlantic Monthly,* December 1998, 85–93.

Brady, Henry E., and David Collier, eds. *Rethinking Social Inquiry: Diverse Tools, Shared Standards.* Lanham, Md.: Rowman & Littlefield, 2004.

Brodie, Bernard, ed., *The Absolute Weapon: Atomic Power and World Order.* New York: Harcourt Brace, 1946.

——. *Strategy in the Missile Age.* Princeton: Princeton University Press, 1959.

——. "Technological Change, Strategic Doctrine, and Political Outcomes." In *Historical Dimensions of National Security Problems,* edited by Klaus Knorr, 263–306. Lawrence: University Press of Kansas, 1976.

Brodie, Bernard, and Fawn M. Brodie. *From Crossbow to H-Bomb.* Bloomington: Indiana University Press, 1973.

Brooks, Richard. "The Italian Campaign of 1859." *Military History* 16 (June 1999).
Brooks, Stephen. "Dueling Realisms." *International Organization* 51 (Summer 1997): 445–78.
Brown, Michael E., Owen R. Coté Jr., Sean M. Lynn-Jones, and Steven E. Miller, eds. *Offense, Defense, and War.* Cambridge: MIT Press, 2004.
Brown, Michael E., Sean M. Lynn-Jones, and Steven E. Miller, eds. *The Perils of An-archy: Contemporary Realism and International Security.* Cambridge: MIT Press, 1995.
Bucholz, Arden. *Moltke and the German Wars, 1864–1871.* New York: Palgrave, 2001.
———. *Moltke, Schlieffen, and Prussian War Planning.* New York: Berg, 1991.
Buhite, Russell D., and Wm. Christopher Hamel. "War for Peace: The Question of an American Preventive War against the Soviet Union, 1945–1955." *Diplomatic History* 14 (Summer 1990): 367–84.
Bundy, McGeorge. "To Cap the Volcano." *Foreign Affairs* 48 (October 1969): 1–20.
Burr, William, ed. "First Strike Options and the Berlin Crisis, September 1961: New Documents from the Kennedy Administration." In *National Security Archive Electronic Briefing Book 56.* National Security Archive, September 25, 2001. http://www.gwu.edu/nsarchiv/nsaebb/nsaebb56.
———. "U.S. Planning for War in Europe, 1963–64." In *National Security Archive Electronic Briefing Book 31.* National Security Archive, May 24, 2000. http://www.gwu.edu/nsarchiv/nsaebb/nsaebb31/index.html.
———. "Missile Defense Thirty Years Ago: Déjà Vu All Over Again?" In *National Security Archive Electronic Briefing Book 36.* National Security Archive, December 18, 2000. http://www.gwu.edu/nsarchiv/NSAEBB/NSAEBB36.
Burr, William, and Jeffrey T. Richelson. "Whether to 'Strangle the Baby in the Cradle': The United States and the Chinese Nuclear Program, 1960–64." *International Security* 25 (Winter 2000–2001): 54–99.
Burt, Richard. *New Weapons Technologies: Debate and Directions, Adelphi Papers.* London: International Institute for Strategic Studies, 1976.
Butfoy, Andrew. "Offence-Defence Theory and the Security Dilemma: The Problem with Marginalizing the Context." *Contemporary Security Policy* 18 (December 1997): 38–58.
Butterfield, Herbert. *History and Human Relations.* London: Collins, 1951.
Calvocoressi, Peter, Guy Wint, and John Pritchard. *Total War: The Causes and Courses of the Second World War.* Rev. 2nd ed. New York: Pantheon Books, 1989.
Camby, Steven. *The Alliance and Europe: Part IV Military Doctrine and Technology, Adelphi Papers.* London: International Institute for Strategic Studies, 1974.
Campbell, E. G. "Railroads in National Defense, 1829–1848." *Mississippi Valley Historical Review* 27 (December 1940): 361–78.
Carr, William. *A History of Germany, 1815–1990.* 4th ed. London: Edward Arnold, 1991.
———. *The Origins of the Wars of German Unification.* London: Longman, 1991.
Christensen, Thomas J. "China, the U.S.-Japan Alliance, and the Security Dilemma in East Asia." *International Security* 23 (Spring 1999): 49–80.

———. "Perceptions and Alliances in Europe, 1865–1940." *International Organization* 51 (Winter 1997): 65–97.

Christensen, Thomas J., and Jack Snyder. "Chain Gangs and Passed Bucks: Predicting Alliance Patterns in Multipolarity." *International Organization* 44 (Spring 1990): 137–68.

Clark, Alan. *Barbarossa: The Russian-German Conflict, 1941–45*. New York: William Morrow, 1965.

Clark, John E. Jr. *Railroads in the Civil War: The Impact of Management on Victory and Defeat*. Baton Rouge: Louisiana State University Press, 2001.

Clausewitz, Carl von. *On War*. Edited and translated by Michael Howard and Peter Paret. Princeton: Princeton University Press, 1976.

Cohen, Eliot A. "A Revolution in Warfare." *Foreign Affairs* 75 (March/April 1996): 37–54.

Collier, David. "Translating Quantitative Methods for Qualitative Researchers: The Case of Selection Bias." *American Political Science Review* 89 (June 1995).

Collier, David, and James Mahoney. "Insights and Pitfalls: Selection Bias in Qualitative Research." *World Politics* 49 (October 1996): 56–91.

Collier, David, James Mahoney, and Jason Seawright. "Claiming Too Much: Warnings about Selection Bias." In *Rethinking Social Inquiry*, edited by Henry E. Brady and David Collier, 85–102. Lanham, Md.: Rowman & Littlefield, 2004.

Cooper, Matthew. *The German Army, 1933–1945: Its Political and Military Failure*. New York: Stein and Day, 1978.

Coox, Alvin D. *Nomonhan: Japan Against Russia, 1939*. 2 vols. Stanford: Stanford University Press, 1985.

Copeland, Dale C. *The Origins of Major War*. Ithaca: Cornell University Press, 2000.

Corum, James S. *The Roots of Blitzkrieg, Hans Von Seeckt and German Military Reform*. Lawrence: University of Kansas, 1992.

Craig, Gordon A. *The Battle of Königgrätz: Prussia's Victory over Austria, 1866*. Philadelphia: J. B. Lippincott, 1964.

Crozier, Brian. *The Rise and Fall of the Soviet Empire*. Rocklin, Calif.: Forum, 1999.

Dastrup, Boyd L. *The Field Artillery: History and Sourcebook*. Westport, Conn.: Greenwood Press, 1994.

Davis, James W., Bernard I. Finel, Stacie E. Goddard, Stephen Van Evera, Charles L. Glaser, and Chaim Kaufmann. "Correspondence: Taking Offense at Offense-Defense Theory." *International Security* 23 (Winter 1998–99): 179–206.

Dean, Jonathan. "Alternative Defense: An Answer to NATO's Central Front Problems?" *International Affairs* 64 (Winter 1987): 61–82.

Dingman, Roger. "Atomic Diplomacy During the Korean War." *International Security* 13 (Winter 1988–89): 50–91.

Dion, Douglas "Evidence and Inference in the Comparative Case Study." *Comparative Politics* 30 (January 1998): 127–45.

Doughty, Robert Allan. *The Breaking Point: Sedan and the Fall of France, 1940*. Hamden, Conn: Archon Books, 1990.

———. *The Seeds of Disaster: The Development of French Army Doctrine, 1919–1939*. Hamden, Conn.: Archon Books, 1985.

Downs, George W., David M. Rocke, and Randolph L. Siverson. "Arms Races and Cooperation." *World Politics* 38 (October 1985): 118–46.

Drea, Edward J. *Nomonhan: Japanese-Soviet Tactical Combat, 1939.* Fort Leavenworth, Kans.: U.S. Army Command and General Staff College, 1981.

Dunnigan, James F. *How to Make War: A Comprehensive Guide to Modern Warfare for the Post-Cold War Era.* 3rd ed. New York: William Morrow, 1988.

Dupuy, R. E., and G. F. Eliot. *If War Comes.* New York: Macmillan, 1937.

Dupuy, R. Ernest, and Trevor N. Dupuy. *The Encyclopedia of Military History.* New York: Harper & Row, 1970.

——. *The Harper Encyclopedia of Military History: From 3500 B.C. to the Present.* New York: HarperCollins, 1993.

Dupuy, T. N. *The Evolution of Weapons and Warfare.* Indianapolis: Bobbs-Merrill, 1980.

——. *Numbers, Predictions, and War: Using History to Evaluate Combat Factors and Predict the Outcome of Battles.* Indianapolis: Bobbs-Merrill, 1979.

——. *Understanding War: History and Theory of Combat.* New York: Paragon House Publishers, 1987.

Eckstein, Harry. "Case Study and Theory in Political Science." Vol. 7, *Handbook of Political Science,* edited by Fred Greenstein and Nelson Polsby, 79–138. Reading, Mass.: Addison-Wesley, 1975.

Edelstein, David M. "Managing Uncertainty: Beliefs about Intentions and the Rise of Great Powers." *Security Studies* 12 (Autumn 2002): 1–40.

Elman, Colin. "Horses for Courses: Why *Not* Neorealist Theories of Foreign Policy?" *Security Studies* 6 (Autumn 1996): 7–53.

Eltis, David. *The Military Revolution in Sixteenth-Century Europe.* London: Tauris Academic Studies, 1995.

Enthoven, Alain C., and K. Wayne Smith. *How Much Is Enough? Shaping the Defense Program, 1961–1969.* New York: Harper & Row, 1971.

Erickson, John. *The Soviet High Command: A Military-Political History, 1918–1941.* London: St. Martin's Press, 1962.

Farrar, Lancelot L. *The Short-War Illusion: German Policy, Strategy and Domestic Affairs, August-December 1914.* Santa Barbara, Calif.: ABC-Clio, 1973.

Fearon, James D. "The Offense-Defense Balance and War since 1648." Paper prepared for the annual meeting of the International Studies Association, Chicago, February 21–25, 1995.

——. "Rationalist Explanations for War." *International Organization* 49 (Summer 1995): 379–414.

Feldman, Shai. *Israeli Nuclear Deterrence: A Strategy for the 1980s.* New York: Columbia University Press, 1982.

——. *Technology and Strategy.* Boulder, Colo.: Westview Press, 1989.

Finel, Bernard I. "Correspondence: Taking Offense at Offense-Defense Theory." *International Security* 23, no. 3 (1998): 182–89.

Fischer, Fritz. *War of Illusions: German Policies from 1911 to 1914.* 1969. Reprint, New York: W. W. Norton, 1975.

Fitzgerald, Frances. *Way Out There in the Blue: Reagan, Star Wars and the End of the Cold War.* New York: Touchstone, 2000.

Flanagan, Stephen J. "Nonprovocative and Civilian-Based Defenses." In *Fateful Visions: Avoiding Nuclear Catastrophe,* edited by Joseph S. Nye, Jr., Graham T. Allison, and Albert Carnesale, 93–109. Cambridge, Mass.: Ballinger, 1988.

Foley, Robert T. "The Origins of the Schlieffen Plan." *War in History* 10 (2003): 222–32.

Förster, Stig. "Dreams and Nightmares: German Military Leadership and the Images of Future Warfare, 1871–1914." In *Anticipating Total War: The German and American Experiences, 1871–1914,* edited by Manfred F. Boemeke, Roger Chickering, and Stig Förster, 343–76. Washington, D.C.: German Historical Institute, 1999.

Förster, Stig and Jörg Nagler, eds. *On the Road to Total War: The American Civil War and the German Wars of Unification, 1861–1871.* Washington, D.C.: German Historical Institute, 1997.

Frankel, Benjamin. *Realism: Restatements and Renewal.* London: Frank Cass, 1996.

——. "Restating the Realist Case: An Introduction." In *Realism: Restatements and Renewal,* edited by Benjamin Frankel, xv–xx. London: Frank Cass, 1996.

Freedman, Lawrence. *The Evolution of Nuclear Strategy.* 3rd ed. Houndmills, UK: Palgrave, 2003.

——. "The First Two Generations of Nuclear Strategists." In *Makers of Modern Strategy: From Machiavelli to the Nuclear Age,* edited by Peter Paret, 735–78. Princeton: Princeton University Press, 1986.

Friedberg, Aaron L. "The Evolution of U.S. Strategic Doctrine, 1945–1980." In *The Strategic Imperative: New Policies for American Security,* edited by Samuel P. Huntington, 53–99. Cambridge, Mass.: Ballinger, 1982.

Friedman, George, and Meredith Friedman. *The Future of War: Power, Technology, and American World Dominance in the 21st Century.* New York: Crown Publishers, 1996.

Fuller, J. F. C. "Aggression and Aggressive Weapons: The Absurdity of Qualitative Disarmament." *Army Ordnance* 14 (1933): 7–11.

——. *Lectures on F.S.R. III.* London: Sifton Praed, 1932.

——. *The Reformation of War.* New York: E. P. Dutton, 1923.

——. *War and Western Civilization, 1832–1932.* London: Duckworth, 1932.

——. "What Is an Aggressive Weapon?" *English Review* 54 (June 1932): 601–5.

Gabel, Christopher R. *Railroad Generalship: Foundations of Civil War Strategy.* Fort Leavenworth, Kans.: U.S. Army Command and General Staff College, 1997.

Gaddis, John Lewis. *The Long Peace: Inquiries into the History of the Cold War.* New York: Oxford University Press, 1989.

Garden, Timothy. *The Technology Trap: Science and the Military.* London: Brassey's Defence Publishers, 1989.

Garthoff, Raymond L. *Detente and Confrontation: American-Soviet Relations from Nixon to Reagan.* Rev. ed. Washington, D.C.: Brookings Institution, 1994.

Gavin, Francis J. "The Myth of Flexible Response: United States Strategy in Europe during the 1960s." *International Historical Review* 23 (December 2001): 847–75.

Geddes, Barbara. "How the Cases You Choose Affect the Answers You Get: Selection Bias in Comparative Politics." In *Political Analysis,* vol. 2, edited by James A. Stimson, 131–50. Ann Arbor: University of Michigan Press, 1990.

George, Alexander L., and Andrew Bennett. *Case Studies and Theory Development in the Social Sciences.* Cambridge: MIT Press, forthcoming.

George, Alexander L., and Timothy J. McKeown. "Case Studies and Theories of Organizational Decision Making." In *Advances in Information Processing in Organizations.* Greenwich, Conn.: JAI Press, 1985.

Gerring, John. "What Is a Case Study and What Is It Good for?" *American Political Science Review* 98 (May 2004): 341–54.

Geyer, Michael. "German Strategy in the Age of Machine Warfare, 1914–1945." In *Makers of Modern Strategy: From Machiavelli to the Nuclear Age,* edited by Peter Paret, 527–97. Princeton: Princeton University Press, 1986.

Gilpin, Robert. *War and Change in World Politics.* Cambridge: Cambridge University Press, 1981.

Glantz, David M. *From the Don to the Dnepr: Soviet Offensive Operations, December 1942–August 1943.* London: Frank Cass, 1991.

Glantz, David M., and Jonathan M. House. *When Titans Clashed: How the Red Army Stopped Hitler.* Lawrence: University Press of Kansas, 1995.

Glaser, Charles L. *Analyzing Strategic Nuclear Policy.* Princeton: Princeton University Press, 1990.

——. "Political Consequences of Military Strategy: Expanding and Refining the Spiral and Deterrence Models." *World Politics* 44 (July 1992): 497–538.

——. "Realists as Optimists: Cooperation as Self-Help." *International Security* 19 (Winter 1994–95): 50–90.

——. "The Security Dilemma Revisited." *World Politics* 50 (October 1997): 171–201.

——. "When Are Arms Races Dangerous? Rational versus Suboptimal Arming." *International Security* 28 (Spring 2004): 44–84.

——. "Why Do Strategists Disagree about the Requirements of Strategic Nuclear Deterrence?" In *Nuclear Arguments: Understanding the Strategic Nuclear Arms and Arms Control Debates,* edited by Lynn Eden and Steven E. Miller, 109–71. Ithaca: Cornell University Press, 1989.

Glaser, Charles L., and Steve Fetter. "National Missile Defense and the Future of U.S. Nuclear Weapons Policy." *International Security* 26 (Summer 2001): 40–92.

Glaser, Charles L., and Chaim Kaufmann. "Does the Offense-Defense Balance Matter?" Paper presented at the annual meeting of the American Political Science Association, Boston, 1998.

——. "What Is the Offense-Defense Balance and Can We Measure It?" *International Security* 22 (Spring 1998): 44–82.

Glynn, Patrick. *Closing Pandora's Box: Arms Races, Arms Control, and the History of the Cold War.* New York: Basic Books, 1992.

Goerlitz, Walter. *History of the German General Staff, 1657–1945.* Boulder, Colo.: Westview Press, 1985.

Goldfischer, David. *The Best Defense: Policy Alternatives for U.S. Nuclear Security from the 1950s to the 1990s.* Ithaca: Cornell University Press, 1993.

Goldman, Emily O., and Richard B. Andres. "Systemic Effects of Military Innovation and Diffusion." *Security Studies* 8 (Summer 1999): 79–125.

Gortzak, Yoav, Yoram Z. Haftel, and Kevin Sweeney. "Offense-Defense Theory: An

Empirical Assessment." *Journal of Conflict Resolution* 49 (February 2005): 67–89.

Gray, Colin S. *House of Cards: Why Arms Control Must Fail.* Ithaca: Cornell University Press, 1992.

——. *Weapons Don't Make War: Policy, Strategy, and Military Technology.* Lawrence: University Press of Kansas, 1993.

Greenwood, Ted. *Making the MIRV: A Study of Defense Decision Making.* Cambridge, Mass.: Ballinger, 1975.

Griffith, Paddy. *Battle Tactics of the Western Front: The British Army's Art of Attack, 1916–18.* New Haven, Conn.: Yale University Press, 1994.

Guderian, Heinz. *Achtung—Panzer!* Translated by Christopher Duffy. London: Arms and Armour Press, 1992.

——. *Panzer Leader.* New York: Da Capo Press, 1996.

Habeck, Mary R. *Storm of Steel: The Development of Armor Doctrine in Germany and the Soviet Union, 1919–1939.* Ithaca: Cornell University Press, 2003.

Hagerman, Edward. *The American Civil War and the Origins of Modern Warfare: Ideas, Organization, and Field Command.* Bloomington: Indiana University Press, 1988.

Hamilton, Richard F., and Holger H. Herwig, eds. *The Origins of World War I.* New York: Cambridge University Press, 2003.

Handel, Michael I. "Clausewitz in the Age of Technology." *Journal of Strategic Studies* 9, nos. 2 and 3 (1986): 51–92.

Harris, J.P. *Men, Ideas, and Tanks: British Military Thought and Armoured Forces, 1903–1939.* Manchester: Manchester University Press, 1995.

——. "The Myth of Blitzkrieg." *War in History* 2 (1995): 335–52.

Harris, J. P., and F. H. Toase. *Armoured Warfare.* New York: St. Martin's Press, 1990.

Haycock, Ronald, and Keith Neilson. *Men, Machines, and War.* Waterloo, Ont.: Wilfrid Laurier University Press, 1988.

Headrick, Daniel. *The Tools of Empire: Technology and European Imperialism in the Nineteenth Century.* New York: Oxford University Press, 1981.

Heidler, David S., and Jeanne T. Heidler, eds. *Encyclopedia of the American Civil War: A Political, Social, and Military History.* New York: W. W. Norton, 2000.

Herrera, Geoffrey Lucas. "The Mobility of Power: Technology, Diffusion, and International System Change." Ph.D. diss., Princeton University, 1995.

Herrmann, David G. *The Arming of Europe and the Making of the First World War.* Princeton: Princeton University Press, 1996.

Herwig, Holger H. "Germany." In *Origins of World War I,* edited by Richard F. Hamilton and Holger H. Herwig, 150–87. New York: Cambridge University Press, 2003.

——. "Germany and the 'Short War' Illusion: Toward a New Interpretation?" *Journal of Military History* 66 (July 2002): 681–94.

Herz, John H. "Idealist Internationalism and the Security Dilemma." *World Politics* 2 (January 1950): 157–80.

Hesolf, Gunilla. "New Technology Favors the Defense." *Bulletin of the Atomic Scientists* 44 (September 1988).

Hillgruber, Andreas. *Germany and the Two World Wars.* Translated by William C. Kirby. Cambridge: Harvard University Press, 1981.

Hoag, Malcolm W. "On Stability in Deterrent Races." *World Politics* 13 (July 1961): 505–27.

Hofmann, George F. "Doctrine, Tank Technology, and Execution: I. A. Khalepskii and the Red Army's Fulfillment of Deep Offensive Operations." *Journal of Slavic Military Studies* 9 (June 1996): 283–334.

Hoffmann, Stanley. *The State of War.* New York: Frederick A. Praeger, 1965.

Holborn, Hajo. "The Prusso-German School: Moltke and the Rise of the General Staff." In *Makers of Modern Strategy from Machiavelli to the Nuclear Age,* edited by Peter Paret, 281–95. Princeton: Princeton University Press, 1986.

Holloway, David. *The Soviet Union and the Arms Race.* New Haven: Yale University Press, 1983.

Holmes, Richard, ed. *The Oxford Companion to Military History.* New York: Cambridge University Press, 2001.

Holmes, Terence M. "Asking Schlieffen: A Further Reply to Terence Zuber." *War in History* 10 (2003): 464–79.

——. "The Real Thing: A Reply to Terence Zuber's 'Terence Holmes Reinvents the Schlieffen Plan.' " *War in History* 9 (2002): 111–20.

——. "The Reluctant March on Paris: A Reply to Terence Zuber's 'The Schlieffen Plan Reconsidered.' " *War in History* 8 (2001): 208–32.

Hopf, Ted. "Polarity, the Offense-Defense Balance, and War." *American Political Science Review* 85 (June 1991): 475–94.

——. "The Promise of Constructivism in International Relations Theory." *International Security* 23 (Summer 1998): 171–200.

House, Jonathan M. *Toward Combined Arms Warfare: A Survey of 20th-Century Tactics, Doctrine, and Organization.* Fort Leavenworth, Kans.: U.S. Army Command and General Staff College, 1984.

Howard, Michael. *The Franco-Prussian War: The German Invasion of France, 1870–1871.* 2nd ed. New York: Macmillan, 1962.

——. "Men against Fire: Expectations of War in 1914." *International Security* 9 (Summer 1984): 41–58.

——. "Review of Offense and Defense in the International System, by George H. Quester." *Survival* 20 (1978): 134–35.

——. *War in European History.* Oxford: Oxford University Press, 1976.

Hughes, B. P. *Firepower: Weapons Effectiveness on the Battlefield, 1630–1850.* London: Arms and Armour Press, 1974.

Hughes, Daniel, ed. *Moltke on the Art of War: Selected Writings.* Novato, Calif.: Presidio, 1993.

Huntington, Samuel P. "U.S. Defense Strategy: The Strategic Innovations of the Reagan Years." In *American Defense Annual, 1987–1988,* edited by Joseph Kruzel, 23–43. Lexington, Mass.: Lexington Books, 1987.

Hybel, Alex Roberto. *The Logic of Surprise in International Conflict.* Lexington, Mass.: Lexington Books, 1986.

India, Government of. "Report of the Kargil Review Committee." New Delhi: Government of India, 2000.

Jackson, Julian. *The Fall of France: The Nazi Invasion of 1940.* New York: Oxford University Press, 2003.

Jacobsen, Carl G. *The Uncertain Course: New Weapons, Strategies and Mind-Sets*. New York: Oxford University Press, 1987.

Jensen, Oliver. *The American Heritage History of Railroads in America*. New York: American Heritage Publishing Co., 1975.

Jervis, Robert. "Cooperation under the Security Dilemma." *World Politics* 30 (January 1978): 167–214.

———. *The Illogic of American Nuclear Strategy*. Ithaca: Cornell University Press, 1984.

———. *The Meaning of the Nuclear Revolution: Statecraft and the Prospect of Armageddon*. Ithaca: Cornell University Press, 1989.

———. *Perception and Misperception in International Politics*. Princeton: Princeton University Press, 1976.

———. *System Effects: Complexity in Political and Social Life*. Princeton: Princeton University Press, 1997.

———. "Was the Cold War a Security Dilemma?" *Journal of Cold War Studies* 3 (Winter 2001): 36–60.

———. "Why Nuclear Superiority Doesn't Matter." *Political Science Quarterly* 94 (Winter 1979–80): 617–33.

Johnson, David E. *Fast Tanks and Heavy Bombers: Innovation in the U.S. Army, 1917–1945*. Ithaca: Cornell University Press, 1998.

Johnson, Hubert C. *Breakthrough!: Tactics, Technology, and the Search for Victory on the Western Front in World War I*. Novato, Calif.: Presidio, 1994.

Johnston, Angus J. II. "Virginia Railroads in April 1861." *Journal of Southern History* 23 (August 1957): 307–30.

Jones, Archer. *The Art of War in the Western World*. New York: Oxford University Press, 1987.

Jones, Matthew. " 'Groping Toward Coexistence': U.S. China Policy during the Johnson Years." *Diplomacy and Statecraft* 12 (September 2001): 175–90.

Kahn, Herman. *On Thermonuclear War*. Princeton: Princeton University Press, 1960.

Kaiser, David. "Deterrence or National Interest? Reflections on the Origins of Wars." *Orbis* 30 (1986).

———. *Politics and War: European Conflict from Philip II to Hitler*. Cambridge: Harvard University Press, 1990.

Kaplan, Fred. "JFK's First-Strike Plan." *Atlantic Monthly*, October 2001, 81–86.

———. *The Wizards of Armageddon*. Stanford: Stanford University Press, 1983.

Katznelson, Ira, and Helen V. Milner, eds. *Political Science: The State of the Discipline*. New York: W. W. Norton, 2002.

Kaufmann, Chaim. "Possible and Impossible Solutions to Ethnic Civil Wars." *International Security* 20 (Spring 1996): 136–75.

Kaufmann, William W. *Military Policy and National Security*. Princeton: Princeton University Press, 1956.

Keefer, Edward. "President Dwight D. Eisenhower and the End of the Korean War." *Diplomatic History* 10 (Summer 1986): 267–89.

Keegan, John. *The First World War*. New York: Alfred A. Knopf, 1999.

———. *A History of Warfare*. New York: Alfred A. Knopf, 1994.

———. *The Second World War*. New York: Penguin Books, 1989.

Kelleher, Catherine M. "Indicators of Defensive Intent in Conventional Force Structures and Operations in Europe." In *Military Power in Europe: Essays in Memory of Jonathan Alford,* edited by Lawrence Freedman, 159–78. New York: St. Martin's Press, 1990.

Kier, Elizabeth. "Culture and Military Doctrine: France between the Wars." *International Security* 19 (Spring 1995): 65–93.

King, Gary, Robert O. Keohane, and Sidney Verba. *Designing Social Inquiry: Scientific Inference in Qualitative Research.* Princeton: Princeton University Press, 1994.

Kissinger, Henry. *Nuclear Weapons and Foreign Policy.* New York: Council on Foreign Relations, 1957.

———. *White House Years.* Boston: Little, Brown, 1979.

Krepinevich, Andrew F. "Cavalry to Computer: The Pattern of Military Revolutions." *National Interest* 37 (Fall 1994): 30–42.

Kydd, Andrew. "Sheep in Sheep's Clothing: Why Security Seekers Do Not Fight Each Other." *Security Studies* 7 (Autumn 1997): 114–55.

Labs, Eric J. "Beyond Victory: Offensive Realism and the Expansion of War Aims." *Security Studies* 6 (Summer 1997): 1–49.

Lakatos, Imre. "Falsification and the Methodology of Scientific Research Programmes." In *Criticism and the Growth of Knowledge,* edited by Imre Lakatos and Alan Musgrave. Cambridge: Cambridge University Press, 1970.

Lash, Jeffrey N. *Destroyer of the Iron Horse: General Joseph E. Johnston and Confederate Rail Transport, 1861–1865.* Kent, Ohio: Kent State University Press, 1991.

Laurance, Edward J. *Light Weapons and Intrastate Conflict: Early Warning Factors and Preventive Action.* New York: Carnegie Corp., 1998.

Lebow, Richard Ned. "Windows of Opportunity: Do States Jump through Them?" *International Security* 9 (1984): 147–86.

Lee, William T. "Soviet Nuclear Targeting Strategy." In *Strategic Nuclear Targeting,* edited by Desmond Ball and Jeffrey Richelson. Ithaca: Cornell University Press, 1986.

Leffler, Melvyn P. *A Preponderance of Power: National Security, the Truman Administration, and the Cold War.* Stanford: Stanford University Press, 1992.

Lennon, Alexander T. J., ed. *Contemporary Nuclear Debates: Missile Defense, Arms Control, and Arms Races in the Twenty-First Century, A Washington Quarterly Reader.* Cambridge: MIT Press, 2002.

Levite, Ariel. *Offense and Defense in Israeli Military Doctrine.* Boulder, Colo.: Westview Press, 1989.

Levy, Jack S. "Historical Trends in Great Power War, 1495–1975." *International Studies Quarterly* 26 (1982): 278–300.

———. "The Offensive/Defensive Balance of Military Technology: A Theoretical and Historical Analysis." *International Studies Quarterly* 28 (June 1984): 219–38.

———. "Preferences, Constraints, and Choices in July 1914." *International Security* 15 (Winter 1990–91): 151–86.

Lewis, George N., Theodore A. Postol, and John Pike. "Why National Missile Defense Won't Work." *Scientific American,* August 1999, 37–41.

Liberman, Peter. "Offense-Defense Balance, Interdependence, and War." *Security Studies* 9 (Summer/Fall 1999): 59–91.

Liddell Hart, B. H. "Aggression and the Problem of Weapons." *English Review* 55 (July 1932): 71–78.

——. *History of the First World War.* London: Faber, 1972.

——. *History of the Second World War.* New York: G. P. Putnam's Sons, 1971.

——. *The Memoirs of Captain Liddell Hart.* London: Cassell & Co., 1965.

Lieber, Keir A. "Grasping the Technological Peace: The Offense-Defense Balance and International Security." *International Security* 25 (Summer 2000): 71–104.

——. "Offense-Defense Theory and the Prospects for Peace." Ph.D. diss., University of Chicago, 2000.

Lieber, Keir, and David Edelstein. "Offense-Defense Theory and Alliance Behavior." Paper prepared for presentation at the Revolution in Military Affairs Conference, Naval Postgraduate School, Monterey, Calif., August 26–29, 1996.

Lind, Jennifer M., and Thomas J. Christensen. "Correspondence: Spirals, Security, and Stability in East Asia." *International Security* 24 (Spring 2000): 190–200.

Loza, Dmitriy. *Commanding the Red Army's Sherman Tanks.* Lincoln: University of Nebraska Press, 1996.

Lupfer, Timothy T. *The Dynamics of Doctrine: The Changes in German Tactical Doctrine During the First World War.* Fort Leavenworth, Kans.: U.S. Army Command and General Staff College, 1981.

Luttwak, Edward, and Stuart L. Koehl. *The Dictionary of Modern War.* New York: Gramercy Books, 1998.

Luvaas, Jay. *The Military Legacy of the Civil War: The European Inheritance.* Chicago: University of Chicago Press, 1959.

——. "A Prussian Observer with Lee," *Military Affairs* 21 (Autumn 1957): 105–17.

Lynn, John A. "Tools of War: Instruments, Ideas, and Institutions of Warfare, 1445–1871." Urbana: University of Illinois Press, 1990.

Lynn-Jones, Sean M. "Offense-Defense Theory and Its Critics." *Security Studies* 4 (Summer 1995): 660–91.

——. "Preface." In *Offense, Defense, and War,* edited by Michael E. Brown, Owen R. Coté Jr., Sean M. Lynn-Jones, and Steven E. Miller, xi–xxxviii. Cambridge: MIT Press, 2004.

——. "Realism and America's Rise: A Review Essay," *International Security* 23 (Fall 1998): 157–82.

——. "Realism, Security, and Offense-Defense Theories: The Implications of Alternative Definitions of the Offense-Defense Balance." Paper presented to the annual meeting of the American Political Science Association, Boston, September 3–6, 1998.

Lynn-Jones, Sean M., and Steven E. Miller. "Preface." In *The Perils of Anarchy: Contemporary Realism and International Security,* edited by Michael E. Brown, Sean M. Lynn-Jones and Steven E. Miller, ix–xiii. Cambridge: MIT Press, 1995.

MacKenzie, Donald. *Inventing Accuracy: A Historical Sociology of Nuclear Missile Guidance.* Cambridge: MIT Press, 1990.

Mandelbaum, Michael. *The Nuclear Revolution.* New York: Cambridge University Press, 1981.

May, Ernest R. *Strange Victory: Hitler's Conquest of France.* New York: Hill and Wang, 2000.

McCullough, David. *Truman.* New York: Simon & Schuster, 1992.

McKeown, Timothy J. "Case Studies and the Statistical Worldview: Review of King, Keohane, and Verba's Designing Social Inquiry: Scientific Inference in Qualitative Research." *International Organization* 53 (Winter 1999): 161–90.

McNeil, William H. *The Age of Gunpowder Empires 1450–1800.* Washington, D.C.: American Historical Association, 1989.

———. *The Pursuit of Power: Technology, Armed Force, and Society since A.D. 1000.* Chicago: University of Chicago Press, 1982.

McPherson, James M. *Battle Cry of Freedom: The Civil War Era.* New York: Ballantine Books, 1988.

Mearsheimer, John J. *Conventional Deterrence.* Ithaca: Cornell University Press, 1983.

———. "The False Promise of International Institutions." *International Security* 19 (1994): 5–49.

———. *Liddell Hart and the Weight of History.* Ithaca: Cornell University Press, 1988.

———. *The Tragedy of Great Power Politics.* New York: W. W. Norton, 2001.

Meredith, Roy and Arthur Meredith. *Mr. Lincoln's Military Railroads.* New York: W. W. Norton, 1979.

Merriman, John. *A History of Modern Europe: From the French Revolution to the Present.* New York: W. W. Norton, 1996.

Miller, David. *The Cold War: A Military History.* New York: St. Martin's Press, 1998.

Miller, Steven E. "Technology and War." *Bulletin of the Atomic Scientists* 41 (1985): 46–48.

Miller, Steven E., Sean M. Lynn-Jones, and Stephen Van Evera. *Military Strategy and the Origins of the First World War.* Rev. and expanded ed. Princeton: Princeton University Press, 1991.

Mlyn, Eric. *The State, Society, and Limited Nuclear War.* Albany: SUNY Press, 1995.

Mombauer, Annika. *Helmuth Von Moltke and the Origins of the First World War.* Cambridge: Cambridge University Press, 2001.

———. *The Origins of the First World War: Controversies and Consensus.* London: Longman, 2002.

Morgan, Patrick M. *Deterrence Now.* Cambridge: Cambridge University Press, 2003.

Morrow, James D. "International Conflict: Assessing the Democratic Peace and Offense-Defense Theory." In *Political Science: The State of the Discipline,* edited by Ira Katznelson and Helen V. Milner, 172–96. New York: W. W. Norton, 2002.

Møller, Bjørn. *Common Security and Nonoffensive Defense: A Neorealist Perspective.* Boulder, Colo.: Lynne Rienner, 1992.

Mueller, John. "The Essential Irrelevance of Nuclear Weapons: Stability in the Postwar World." *International Security* 13 (Autumn 1988): 55–79.

Mueller, Karl. "Alignment Balancing and Stability in Eastern Europe." *Security Studies* 5 (Autumn 1995): 38–76.

Murray, Williamson. "Armored Warfare: The British, French, and German Experiences." In *Military Innovation in the Interwar Period,* edited by Williamson Murray and Allan R. Millett. Cambridge: Cambridge University Press, 1996.

Murray, Williamson, and Alan R. Millett, eds. *Military Innovation in the Interwar Period.* Cambridge: Cambridge University Press, 1996.

——. *A War to Be Won: Fighting the Second World War.* Cambridge, Mass.: Belknap, 2000.

Natural Resources Defense Council. *Archive of Nuclear Data.* http://www.nrdc.org/nuclear/nudb/datainx.asp.

Nef, John U. *War and Human Progress: An Essay on the Rise of Industrial Civilization.* New York: W. W. Norton, 1963.

Nitze, Paul H. "Assuring Strategic Stability in an Era of Détente." *Foreign Affairs* 54 (January 1976): 207–32.

Norris, Robert S., and Thomas B. Cochran. "U.S. and USSR/Russian Strategic Offensive Nuclear Forces, 1945–1966." Washington, D.C.: Natural Resources Defense Council, 1997.

Nye, Joseph S. Jr., Graham T. Allison and Albert Carnesale, eds. *Fateful Visions: Avoiding Nuclear Catastrophe.* Cambridge, Mass.: Ballinger, 1988.

O'Connell, Robert. *Of Arms and Men: A History of War, Weapons, and Aggression.* New York: Oxford University Press, 1989.

Ogorkiewicz, R. M. *Armoured Forces: A History of Armoured Forces and Their Vehicles.* New York: Arco, 1970.

Osgood, Robert E., and Robert C. Tucker. *Force, Order, and Justice.* Baltimore: Johns Hopkins University Press, 1967.

Pakenham, Thomas. *The Boer War.* New York: Random House, 1994.

Parker, Geoffrey. *The Military Revolution: Military Innovation and the Rise of the West, 1500–1800.* 2nd ed. Cambridge: Cambridge University Press, 1996.

Paul, T. V., Richard J. Harknett, and James J. Wirtz. *The Absolute Weapon Revisited: Nuclear Arms and the Emerging International Order.* Ann Arbor: University of Michigan Press, 1998.

Payne, Keith B. *The Fallacies of Cold War Deterrence and a New Direction.* Lexington: The University Press of Kentucky, 2001.

Pearton, Maurice. *Diplomacy, War, and Technology since 1830.* Lawrence: University Press of Kansas, 1984.

——. *The Knowledgeable State: Diplomacy, War and Technology since 1830.* London: Burnett Books, 1982.

Pickenpaugh, Roger. *Rescue by Rail: Troop Transfer and the Civil War in the West, 1863.* Lincoln: University of Nebraska Press, 1998.

Pike, John. *Nuclear Forces Guide,* 1999. http://www.fas.org/nuke/guide/summary.htm.

Podvig, Pavel, ed. *Russian Strategic Nuclear Forces.* Cambridge: MIT Press, 2001.

Porch, Douglas. *The March to the Marne: The French Army, 1871–1914.* Cambridge: Cambridge University Press, 1981.

Porter, Bruce D. "Washington, Moscow, and Third World Conflict in the 1980s." In *The Strategic Imperative: New Policies for American Security,* edited by Samuel P. Huntington. Cambridge, Mass.: Ballinger, 1982.

Posen, Barry R. *Inadvertent Escalation: Conventional War and Nuclear Risks.* Ithaca: Cornell University Press, 1991.

——. "The Security Dilemma and Ethnic Conflict." *Survival* 35 (Spring 1993): 27–57.

——. *The Sources of Military Doctrine: France, Britain, and Germany between the World Wars.* Ithaca: Cornell University Press, 1984.

Posen, Barry R., and Stephen Van Evera. "Defense Policy and the Reagan Administration: Departure from Containment." *International Security* 8 (Summer 1983): 3–45.

Powell, Robert. "Guns, Butter, and Anarchy." *American Political Science Review* 87 (March 1993): 115–32.

——. *In the Shadow of Power.* Princeton: Princeton University Press, 1999.

Pratt, Edwin A. *The Rise of Rail-Power in War and Conquest, 1833–1914.* Philadelphia: J. B. Lippincott, 1916.

Press, Daryl G. *Calculating Credibility: How Leaders Assess Military Threats.* Ithaca: Cornell University Press, 2005.

Quester, George H. *Nuclear Monopoly.* New Brunswick, N.J.: Transaction, 2000.

——. *Offense and Defense in the International System.* New York: John Wiley, 1977.

Ramsdell, Charles W. "The Confederate Government and the Railroads." *American Historical Review* 22 (July 1917): 794–810.

Ravenal, Earl C. "Counterforce and Alliance: The Ultimate Connection." *International Security* 6 (Spring 1982): 26–43.

Reid, Brian Holden. *J. F. C. Fuller: Military Thinker.* New York: St. Martin's Press, 1987.

Rhodes, Richard. *Dark Sun: The Making of the Hydrogen Bomb.* New York: Touchstone, 1995.

——. *The Making of the Atomic Bomb.* New York: Touchstone, 1988.

Ritter, Gerhard. *The Schlieffen Plan: Critique of a Myth.* New York: Praeger, 1958.

——. *The Sword and the Scepter: The Problem of Militarism in Germany.* 4 vols. Coral Gables, Fla.: University of Miami Press, 1969.

Rogowski, Ronald. "The Role of Theory and Anomaly in Social-Scientific Inference." *American Political Science Review* 89 (June 1995): 467–70.

Ropp, Theodore. *War in the Modern World.* New York: Collier Books, 1962.

Rose, Gideon. "Neoclassical Realism and Theories of Foreign Policy." *World Politics* 51 (October 1998): 144–72.

Rosenberg, David. "Reality and Responsibility: Power and Process in the Making of United States Nuclear Strategy, 1945–1968." *Journal of Strategic Studies* 9 (March 1986): 35–52.

——. "The Origins of Overkill: Nuclear Weapons and American Strategy, 1945–1960." *International Security* 7 (Spring 1983): 3–71.

——. " 'A Smoking Radiating Ruin at the End of Two Hours': Documents on American Plans for Nuclear War with the Soviet Union, 1954–1955." *International Security* 6 (Winter 1981–82): 3–38.

Rothenberg, Gunther E. "Moltke, Schlieffen, and the Doctrine of Strategic Envelopment." In *Makers of Modern Strategy from Machiavelli to the Nuclear Age,* edited by Peter Paret, 296–325. Princeton: Princeton University Press, 1986.

Royle, Trevor. *Crimea: The Great Crimean War, 1854–1856.* New York: St. Martin's Press, 2000.

Rueschemeyer, Dietrich. "Can One or a Few Cases Yield Theoretical Gains?" In

Comparative Historical Analysis in the Social Sciences, edited by James Mahoney and Dietrich Rueschemeyer, 305–36. New York: Cambridge University Press, 2003.

Russell, Richard L. "Persian Gulf Proving Grounds: Testing Offense-Defence Theory." *Contemporary Security Policy* 23 (December 2002): 192–213.

Sagan, Scott D. *Moving Targets: Nuclear Strategy and National Security.* Princeton: Princeton University Press, 1989.

———. "1914 Revisited: Allies, Offense, and Instability." *International Security* 11 (Fall 1986): 151–75.

———. "SIOP-62: The Nuclear War Plan Briefing to President Kennedy." *International Security* 12 (Summer 1987): 22–51.

Sagan, Scott D., and Kenneth N. Waltz, *The Spread of Nuclear Weapons: A Debate Renewed.* New York: W. W. Norton, 2003.

Sagle, Lawrence W. "Railroads in Wartime, 1861–1865." Unpublished manuscript, Towson, Md., 1965.

Salman, Michael, Kevin J. Sullivan, and Stephen Van Evera. "Analysis or Propaganda? Measuring American Strategic Nuclear Capability, 1969–88." In *Nuclear Arguments: Understanding the Strategic Nuclear Arms and Arms Control Debates,* edited by Lynn Eden and Steven E. Miller, 172–263. Ithaca: Cornell University Press, 1989.

Saveliyev, Aleksandr G., and Nikolay N. Detinov. *The Big Five: Arms Control Decision-Making in the Soviet Union.* Translated by Dimitryi Trenin. Edited by Gregory Vorhall. Westport, Conn.: Praeger, 1995.

Scales, Robert H., Jr. *America's Army in Transition: Preparing for War in the Precision Age.* Carlisle, Pa.: U.S. Army War College, 1999.

Schelling, Thomas C. *Arms and Influence.* New Haven: Yale University Press, 1966.

———. *The Strategy of Conflict.* New York: Oxford University Press, 1963.

Schelling, Thomas C., and Morton H. Halperin. *Strategy and Arms Control.* New York: Twentieth Century Fund, 1961.

Schwartz, Stephen I., ed. *Atomic Audit: The Costs and Consequences of U.S. Nuclear Weapons since 1940.* Washington, D.C.: Brookings Institution, 1998.

———. *How Much Was Enough? Official Estimates of Nuclear Weapons Requirements, 1957–95.* 1998. http://www.brook.edu/fp/projects/nucwcost/figure3.htm.

Schweller, Randall L. "Bandwagoning for Profit: Bringing the Revisionist State Back In," *International Security* 19 (Summer 1994): 72–107.

———. *Deadly Imbalances: Tripolarity and Hitler's Strategy of World Conquest.* New York: Columbia University Press, 1998.

———. "Neorealism's Status-Quo Bias: What Security Dilemma?" *International Security* 5 (Spring 1996): 90–121.

Seaton, Albert. *The Fall of Fortress Europe, 1943–1945.* New York: Holmes & Meier, 1981.

"Select Enemy. Delete." *Economist,* March 8, 1997, 21–24.

Sherman, William T. "French Mobilization in 1870." In *On the Road to Total War: The American Civil War and the German Wars of Unification, 1861–1871,* edited by Stig Förster and Jörg Nagler, 283–94. Washington, D.C.: German Historical Institute, 1997.

——. *Memoirs of General W. T. Sherman.* Vol. 2. New York: Charles L. Webster, 1891.

Shimshoni, Jonathan. "Technology, Military Advantage and World War I: A Case for Military Entrepreneurship." *International Security* 15 (Winter 1990–91): 187–215.

Showalter, Dennis. "Mass Multiplied by Impulsion: The Influence of Railroads on Prussian Planning for the Seven Weeks' War." *Military Affairs* 38, no. 2 (1974): 62–67.

——. "Prussia, Technology, and War: Artillery from 1815 to 1914." In *Men, Machines, and War,* edited by Ronald Haycock and Keith Neilson, 113–51. Waterloo, Ont.: Wilfried Laurier University Press, 1988.

——. *Railroads and Rifles: Soldiers, Technology, and the Unification of Germany.* Hamden, Conn.: Archon Books, 1975.

——. *Tannenberg: Clash of Empires.* Hamden, Conn.: Archon Books, 1991.

Snyder, Glenn H. "The Balance of Power and the Balance of Terror." In *The Balance of Power,* edited by Paul Seabury, 184–201. San Francisco: Chandler, 1965.

——. *Deterrence and Defense: Toward a Theory of National Security.* Princeton: Princeton University Press, 1961.

——. "The Security Dilemma in Alliance Politics." *World Politics* 36 (1984): 461–95.

Snyder, Jack. "Civil-Military Relations and the Cult of the Offensive, 1914 and 1984." *International Security* 9 (Summer 1984): 108–46.

——. *The Ideology of the Offensive: Military Decision Making and the Disasters of 1914.* Ithaca: Cornell University Press, 1984.

——. *Myths of Empire: Domestic Politics and International Ambition.* Ithaca: Cornell University Press, 1991.

Spielvogel, Jackson J. *Hitler and Nazi Germany: A History.* Englewood Cliffs, N.J.: Prentice Hall, 1988.

Stevenson, David. *Armaments and the Coming of War: Europe, 1904–1914.* Oxford: Oxford University Press, 1996.

——. *Cataclysm: The First World War as Political Tragedy.* New York: Basic Books, 2004.

——. "Militarization and Diplomacy in Europe before 1914." *International Security* 22 (Summer 1997): 125–61.

——. "War by Timetable? The Railway Race Before 1914." *Past and Present* 162 (February 1999): 163–94.

Stone, Norman. *The Eastern Front, 1914–1917.* New York: Charles Scribner's Sons, 1975.

Strachan, Hew. *European Armies and the Conduct of War.* London: Routledge, 1983.

——. *The First World War.* New York: Viking, 2003.

——. *The First World War: To Arms.* Vol. 1. Oxford: Oxford University Press, 2001.

Symonds, Craig L. *A Battlefield Atlas of the Civil War.* Baltimore: Nautical and Aviation Publishing Co. of America, 1983.

Taliaferro, Jeffrey, V. "Security Seeking under Anarchy: Defensive Realism Revisited." *International Security* 25 (Winter 2000–2001): 128–61.

Tate, Merze. *The United States and Armaments.* Cambridge: Harvard University Press, 1948.

Taylor, A. J. P. *The First World War: An Illustrated History.* 1963. Reprint, New York: Perigree Books, 1980.

——. *War by Time-Table: How the First World War Began.* London: Macdonald, 1969.

Townshend, Charles. "Introduction: The Shape of Modern War." In *The Oxford Illustrated History of Modern War,* edited by Charles Townshend, 3–18. Oxford: Oxford University Press, 1997.

Trachtenberg, Marc. *History and Strategy.* Princeton: Princeton University Press, 1991.

——. "The Meaning of Mobilization in 1914." *International Security* 15 (Winter 1990–91): 120–50.

——. "A 'Wasting Asset': American Strategy and the Shifting Nuclear Balance, 1949–1954," *International Security* 13 (Winter 1988–89): 5–48.

Travers, T. H. E. "Technology, Tactics, and Morale: Jean De Bloch, the Boer War, and British Military Theory, 1900–1914." *Journal of Modern History* 51 (June 1979): 264–86.

——. *How the War Was Won: Command and Technology in the British Army on the Western Front, 1917–1918.* New York: Routledge, 1992.

——. *The Killing Ground: The British Army, the Western Front and the Emergence of Modern Warfare, 1900–1918.* London: Allen & Unwin, 1987.

Tuchman, Barbara W. *The Guns of August.* New York: Macmillan, 1962.

Tucker, Spencer C. *The Great War, 1914–18.* Bloomington: Indiana University Press, 1998.

Turnbull, Patrick. *Solferino: The Birth of a Nation.* New York: St. Martin's Press, 1985.

Turner, George Edgar. *Victory Rode the Rails: The Strategic Place of the Railroads in the Civil War.* Lincoln: University of Nebraska Press, 1992.

Turner, L. C. F. *Origins of the First World War.* New York: W. W. Norton, 1970.

United States Department of State. *Foreign Relations of the United States.* 1946. Vol. 1: General. Washington, D.C.: GPO, 1972.

——. *Foreign Relations of the United States.* 1948. Vol. 1: General. Washington, D.C.: GPO, 1975.

——. *Foreign Relations of the United States.* 1949. Vol. 1: National Security Affairs. Washington, D.C.: GPO, 1976.

——. *Foreign Relations of the United States.* 1950. Vol. 1: National Security Affairs. Washington, D.C.: GPO, 1977.

——. *Foreign Relations of the United States.* 1952–54. Vol. 2: National Security Affairs. Washington, D.C.: GPO, 1984.

——. *Foreign Relations of the United States.* 1961–63. Vol. 8: National Security Policy. Washington, D.C.: GPO, 1996.

Van Creveld, Martin. *Supplying War: Logistics from Wallenstein to Patton.* Cambridge: Cambridge University Press, 1977.

——. *Technology and War: From 2000 B.C. to the Present.* New York: Free Press, 1991.

Van Evera, Stephen. *Causes of War: Power and the Roots of Conflict.* Ithaca: Cornell University Press, 1999.

——. "The Cult of the Offensive and the Origins of the First World War." *International Security* 9 (Summer 1984): 58–107.

———. *Guide to Methods for Students of Political Science.* Ithaca: Cornell University Press, 1997.

———. "Offense, Defense, and the Causes of War." *International Security* 22 (Spring 1998): 5–43.

———. "Taking Offense at Offense-Defense Theory." *International Security* 23, no. 3 (1998): 195–200.

Von Manstein, Erich. *Lost Victories.* Edited and translated by Anthony G. Powell. Novato, Calif.: Presidio Press, 1982.

Von Mellenthin, F. W. *Panzer Battles: A Study of the Employment of Armor in the Second World War.* New York: Ballantine Books, 1956.

Von Moltke, Helmuth. "Defensive and Offensive (1874)." In *Moltke on the Art of War: Selected Writings,* edited by Daniel Hughes. Novato, Calif.: Presidio, 1993.

———. "Defensive Position, Envelopment, and Base of Operations (1865)." In *Moltke on the Art of War: Selected Writings,* edited by Daniel Hughes. Novato, Calif.: Presidio, 1993.

———. "Instructions for Large Unit Commanders of 24 June 1869." In *Moltke on the Art of War: Selected Writings,* edited by Daniel Hughes. Novato, Calif.: Presidio, 1993.

Wagner, Margaret E., Gary W. Gallagher, and Paul Finkelman, eds. *The Library of Congress Civil War Desk Reference.* New York: Simon & Schuster, 2002.

Waller, Bruce. *Bismarck.* 2nd ed. New York: Blackwell, 1997.

Walt, Stephen M. "The Case for Finite Containment: Analyzing U.S. Grand Strategy." *International Security* 14 (Summer 1989): 5–49.

———. "The Enduring Relevance of the Realist Tradition." In *Political Science: The State of the Discipline,* edited by Ira Katznelson and Helen V. Milner, 197–234. New York: W. W. Norton, 2002.

———. *The Origins of Alliances.* Ithaca: Cornell University Press, 1987.

———. *Revolution and War.* Ithaca: Cornell University Press, 1996.

———. "Revolution and War." *World Politics* 44 (April 1992): 321–68.

Walter, Barbara F., and Jack Snyder. *Civil Wars, Insecurity, and Intervention.* New York: Columbia University Press, 1999.

Waltz, Kenneth N. "Nuclear Myths and Political Realities." *American Political Science Review* 84 (September 1990): 731–45.

———. *The Spread of Nuclear Weapons: More May Be Better.* Adelphi Papers no. 171. London: International Institute for Strategic Studies, 1981.

———. *Theory of International Politics.* New York: McGraw-Hill, Inc., 1979.

Ward, James A. *Railroads and the Character of America, 1820–1887.* Knoxville: University of Tennessee Press, 1986.

Watt, Donald Cameron. *How War Came: The Immediate Origins of the Second World War, 1938–1939.* London: Pimlico, 2001.

Watts, Barry, and Williamson Murray. "Military Innovation in Peacetime." In *Military Innovation in the Interwar Period,* edited by Williamson Murray and Allan R. Millett, 369–415. Cambridge: Cambridge University Press, 1996.

Wawro, Geoffrey. *The Austro-Prussian War: Austria's War with Prussia and Italy in 1866.* New York: Cambridge University Press, 1996.

——. *The Franco-Prussian War: The German Conquest of France in 1870–1871.* New York: Cambridge University Press, 2003.

Weber, Thomas. *The Northern Railroads in the Civil War, 1861–1865.* Bloomington: Indiana University Press, 1952.

Weigley, Russell F. *A Great Civil War: A Military and Political History, 1861–1865.* Bloomington: Indiana University Press, 2000.

Weinberg, Gerhard L. *The Foreign Policy of Hitler's Germany: Diplomatic Revolution in Europe, 1933–36.* Chicago: University of Chicago Press, 1970.

——. *The Foreign Policy of Hitler's Germany: Starting World War II, 1937–1939.* Chicago: University of Chicago Press, 1980.

Wendt, Alexander. "Anarchy Is What States Make of It: The Social Construction of Power Politics." *International Organization* 46 (Spring 1992): 391–425.

——. *Social Theory of International Politics.* New York: Cambridge University Press, 1999.

Westwood, John. *Railways at War.* San Diego: Howell-North Books, 1980.

Willmott, H. P., and John Keegan. *The Second World War in the East.* London: Cassell, 1999.

Wiseman, Geoffrey. *Concepts of Non-Provocative Defence: Ideas and Practices in International Security.* New York: Palgrave, 2002.

Wohlstetter, Albert. "The Delicate Balance of Terror." *Foreign Affairs* 37 (1959): 211–34.

——. "The Political and Military Aims of Offense and Defense Innovation." In *Swords and Shields,* edited by Fred S. Hoffman, Albert Wohlstetter, and David S. Yost. Lexington, Mass.: D.C. Heath, 1987.

Wray, Timothy A. *Standing Fast: German Defensive Doctrine on the Russian Front During World War II, Prewar to March 1943.* Combat Studies Institute, Research Survey. Fort Leavenworth, Kans: U.S. Army Command and General Staff College, 1986.

Wright, Quincy. *A Study of War.* Chicago: University of Chicago Press, 1965.

Yin, Robert K. *Case Study Research: Design and Methods.* 2nd ed. Thousand Oaks, Calif.: Sage, 1994.

Zakaria, Fareed. "Realism and Domestic Politics." *International Security* 17 (1992): 177–98.

——. *From Wealth to Power: The Unusual Origins of America's World Role.* Princeton: Princeton University Press, 1998.

Zaloga, Steven J. *The Kremlin's Nuclear Sword: The Rise and Fall of Russia's Strategic Nuclear Forces, 1945–2000.* Washington, D.C.: Smithsonian Institution Press, 2002.

Zuber, Terence. *Inventing the Schlieffen Plan: German War Planning, 1871–1914.* New York: Oxford University Press, 2002.

——. "The Schlieffen Plan Reconsidered." *War in History* 6 (1999): 262–305.

——. "Terence Holmes Reinvents the Schlieffen Plan." *War in History* 8 (2001): 468–76.

——. "Terence Holmes Reinvents the Schlieffen Plan—Again." *War in History* 10 (2003): 92–101.

Notes

Introduction

1. For example, on the role of technology in the rise of the West see William H. Mc-Neil, *The Pursuit of Power: Technology, Armed Force, and Society since A.D. 1000* (Chicago: University of Chicago Press, 1982); Maurice Pearton, *Diplomacy, War, and Technology since 1830* (Lawrence: University Press of Kansas, 1984); Geoffrey Parker, *The Military Revolution: Military Innovation and the Rise of the West, 1500–1800,* 2nd ed. (Cambridge: Cambridge University Press, 1996); and Daniel Headrick, *The Tools of Empire: Technology and European Imperialism in the Nineteenth Century* (New York: Oxford University Press, 1981). On the impact of technology on warfare see Bernard Brodie and Fawn M. Brodie, *From Crossbow to H-Bomb* (Bloomington: Indiana University Press, 1973); and T. N. Dupuy, *The Evolution of Weapons and Warfare* (Indianapolis: Bobbs-Merrill, 1980); Robert O'Connell, *Of Arms and Men: A History of War, Weapons, and Aggression* (New York: Oxford University Press, 1989); Martin van Creveld, *Technology and War: From 2000 B.C. to the Present* (New York: Free Press, 1991).

2. The foundational works on offense-defense theory are Robert Jervis, "Cooperation under the Security Dilemma," *World Politics* 30 (January 1978): 167–214; Stephen Van Evera, *Causes of War: Power and the Roots of Conflict* (Ithaca: Cornell University Press, 1999), especially chap. 6; and George H. Quester, *Offense and Defense in the International System* (New York: John Wiley, 1977). For theoretical developments, refinements, and extensions, see Stephen Van Evera, "Offense, Defense, and the Causes of War," *International Security* 22 (Spring 1998): 5–43; Charles L. Glaser and Chaim Kaufmann, "What Is the Offense-Defense Balance and Can We Measure It?" *International Security* 22 (Spring 1998): 44–82; Sean M. Lynn-Jones, "Offense-Defense Theory and Its Critics," *Security Studies* 4 (Summer 1995): 660–91; and Charles L. Glaser, "Realists as Optimists:

Cooperation as Self-Help," *International Security* 19 (Winter 1994–95): 50–90. Stephen
Biddle proposes what he calls a new version of offense-defense theory, but it should be
seen as an altogether new theory of military capability. See Stephen Biddle, "Rebuilding the Foundations of Offense-Defense Theory," *Journal of Politics* 63 (August 2001):
741–74. A compilation of many of these works is Michael E. Brown and others, eds., *Offense, Defense, and War* (Cambridge: MIT Press, 2004).

3. Van Evera, *Causes of War,* 117.

4. Jervis, "Cooperation under the Security Dilemma," 191. Also see Stephen Van
Evera, "The Cult of the Offensive and the Origins of the First World War," *International
Security* 9 (Summer 1984): 58–107; Jack Snyder, "Civil-Military Relations and the Cult of
the Offensive, 1914 and 1984," *International Security* 9 (Summer 1984): 108–46; Jack
Snyder, *The Ideology of the Offensive: Military Decision Making and the Disasters of 1914*
(Ithaca: Cornell University Press, 1984); and Van Evera, *Causes of War,* chap. 7.

5. The following works either apply, test, or critique the offense-defense balance.
On interstate war, in addition to the core works cited above, see Ted Hopf, "Polarity,
the Offense-Defense Balance, and War," *American Political Science Review* 85 (June
1991): 475–94; James D. Fearon, "The Offense-Defense Balance and War since 1648"
(paper presented at the annual convention of the International Studies Association,
Chicago, February 21–25, 1995); Andrew Butfoy, "Offence-Defence Theory and the
Security Dilemma: The Problem with Marginalizing the Context," *Contemporary Security Policy* 18 (December 1997): 38–58; Peter Liberman, "The Offense-Defense Balance, Interdependence, and War," *Security Studies* 9 (Summer and Fall 1999): 59–91;
Keir A. Lieber, "Grasping the Technological Peace: The Offense-Defense Balance and
International Security," *International Security* 25 (Summer 2000): 71–104; Stephen D.
Biddle, "Testing Offense-Defense Theory: The Second Battle of the Somme, March 21
to April 9, 1918" (paper presented at the annual meeting of the American Political
Science Association, Atlanta, September 2–5, 1999); Biddle, "Rebuilding the Foundations of Offense-Defense Theory," 741–74; Richard L. Russell, "Persian Gulf Proving
Grounds: Testing Offense-Defence Theory," *Contemporary Security Policy* 23 (December
2002): 192–213; Karen Ruth Adams, "Attack and Conquer? International Anarchy and
the Offense-Defense-Deterrence Balance," *International Security* 28 (Winter 2003–4):
45–83; and Yoav Gortzak, Yoram Z. Haftel, and Kevin Sweeney, "Offense-Defense Theory: An Empirical Assessment," *Journal of Conflict Resolution* 49 (February 2005): 67–
89. On ethnic and civil conflict see Barry R. Posen, "The Security Dilemma and Ethnic
Conflict," *Survival* 35 (Spring 1993): 27–57; Chaim Kaufmann, "Possible and Impossible Solutions to Ethnic Civil Wars," *International Security* 20 (Spring 1996): 136–75;
and Barbara F. Walter and Jack Snyder, eds., *Civil Wars, Insecurity, and Intervention*
(New York: Columbia University Press, 1999). On arms control and arms racing see
Malcolm W. Hoag, "On Stability in Deterrent Races," *World Politics* 13 (July 1961):
505–27; Quester, *Offense and Defense in the International System;* Jervis, "Cooperation
under the Security Dilemma,"; George W. Downs, David M. Rocke, and Randolph M.
Siverson, "Arms Races and Cooperation," *World Politics* 38 (October 1985): 118–46;
Charles L. Glaser, "Political Consequences of Military Strategy: Expanding and Refining the Spiral and Deterrence Models," *World Politics* 44 (July 1992): 497–538; Robert
Powell, "Guns, Butter, and Anarchy," *American Political Science Review* 87 (March 1993):
115–32; Robert Powell, *In the Shadow of Power* (Princeton: Princeton University Press,
1999); and Charles L. Glaser, "When Are Arms Races Dangerous? Rational versus Suboptimal Arming," *International Security* 28 (Spring 2004): 44–84. On alliance behavior

see Stephen M. Walt, *The Origins of Alliances* (Ithaca: Cornell University Press, 1987); Thomas J. Christensen and Jack Snyder, "Chain Gangs and Passed Bucks: Predicting Alliance Patterns in Multipolarity," *International Organization* 44 (Spring 1990): 137–68; Karl Mueller, "Alignment Balancing and Stability in Eastern Europe," *Security Studies* 5 (Autumn 1995): 38–76; and Thomas J. Christensen, "Perceptions and Alliances in Europe, 1865–1940," *International Organization* 51 (Winter 1997): 65–97. On crisis behavior see Thomas C. Schelling, *Arms and Influence* (New Haven: Yale University Press, 1966); and Van Evera, *Causes of War.* On military doctrine, see Van Evera, "Cult of the Offensive"; Jack Snyder, "Civil-Military Relations"; Barry R. Posen, *The Sources of Military Doctrine: France, Britain, and Germany between the World Wars* (Ithaca: Cornell University Press, 1984); Scott D. Sagan, "1914 Revisited: Allies, Offense, and Instability," *International Security* 11 (Fall 1986): 151–75; Jonathan Shimshoni, "Technology, Military Advantage and World War I: A Case for Military Entrepreneurship," *International Security* 15 (Winter 1990–91): 187–215; Elizabeth Kier, "Culture and Military Doctrine: France Between the Wars," *International Security* 19 (Spring 1995): 65–93; and Van Evera, *Causes of War.* On revolutions see Stephen M. Walt, "Revolution and War," *World Politics* 44 (April 1992): 321–68; and Stephen M. Walt, *Revolution and War* (Ithaca: Cornell University Press, 1996). On the number and size of states and empires and international system structure see Stanislav Andreski, *Military Organization and Society* (London: Routledge and Keegan Paul, 1968), 75–76; Richard Bean, "War and the Birth of the Nation State," *Journal of Economic History* 33 (March 1973): 207–21; Quester, *Offense and Defense in the International System;* and Robert Gilpin, *War and Change in World Politics* (Cambridge: Cambridge University Press, 1981), 61–62. On grand strategy, see Stephen M. Walt, "The Case for Finite Containment: Analyzing U.S. Grand Strategy," *International Security* 14 (Summer 1989): 5–49; and Barry R. Posen, *Inadvertent Escalation: Conventional War and Nuclear Risks* (Ithaca: Cornell University Press, 1991).

6. Van Evera, *Causes of War,* 185.

7. On whether intentions can be effectively conveyed, see Glaser, "When Are Arms Races Dangerous?" 55–58; Glaser, "Realists as Optimists"; Andrew Kydd, "Sheep in Sheep's Clothing: Why Security Seekers Do Not Fight Each Other," *Security Studies* 7 (Autumn 1997): 114–55; Randall L. Schweller, "Neorealism's Status-Quo Bias: What Security Dilemma?" *Security Studies* 5 (Spring 1996): 90–121; and David M. Edelstein, "Managing Uncertainty: Beliefs about Intentions and the Rise of Great Powers," *Security Studies* 12 (Autumn 2002): 1–40. Offense-defense proponents note, however, that even if total differentiation of offensive and defensive forces were possible, status quo states might still need to build offense when the offense-defense balance favors offense or when they need to protect distant allies, retake territory lost at the outset of a conflict, or bolster deterrence. Jervis, "Cooperation under the Security Dilemma," 201–2; Charles L. Glaser, "The Security Dilemma Revisited," *World Politics* 50 (October 1997):186n; Van Evera, *Causes of War,* 152–60.

8. The argument that offensive and defensive weapons cannot be distinguished has been the most common criticism of offense-defense theory. See John J. Mearsheimer, *Conventional Deterrence* (Ithaca: Cornell University Press, 1983), 24–27; Samuel P. Huntington, "U.S. Defense Strategy: The Strategic Innovations of the Reagan Years," in *American Defense Annual, 1987–1988,* ed. Joseph Kruzel(Lexington, Mass.: Lexington Books, 1987), 23–43, at 35–37; Colin S. Gray, *House of Cards: Why Arms Control Must Fail* (Ithaca: Cornell University Press, 1992), 28, 66–68; Colin S. Gray, *Weapons Don't Make*

War: Policy, Strategy, and Military Technology (Lawrence: University of Kansas Press, 1993), chap. 2.

9. See, for example, Van Evera, *Causes of War,* 177.

10. I thank Robert Jervis for suggesting this example.

11. The basic elements of offensive realism are discussed at greater length below. See John J. Mearsheimer, *The Tragedy of Great Power Politics* (New York: W. W. Norton, 2001).

12. Richard K. Betts, "Must War Find a Way? A Review Essay," *International Security* 24 (Fall 1999): 166–98.

13. Melvyn P. Leffler, *A Preponderance of Power: National Security, the Truman Administration, and the Cold War* (Stanford: Stanford University Press, 1992), 327–33; Richard Rhodes, *The Making of the Atomic Bomb* (New York: Touchstone, 1988), 768; Richard Rhodes, *Dark Sun: The Making of the Hydrogen Bomb* (New York: Touchstone, 1995).

14. U. S. Department of State, *Report of the Special Committee of the NSC to Truman,* January 31, 1950, in *Foreign Relations of the United States* (1950), 1:515. As David McCullough recounts President Truman's final discussion with his top advisers about pursuing the hydrogen bomb: " 'Can the Russians do it?' he asked the group. It was his only question. They all nodded. 'We don't have much time,' interjected Admiral Souers. 'In that case,' said Truman, 'we have no choice. We'll go ahead.' " David McCullough, *Truman* (New York: Simon & Schuster, 1992), 763.

15. Carl von Clausewitz, *On War,* ed. and trans. Michael Howard and Peter Paret (Princeton: Princeton University Press, 1976), 217–18, 359. On offense-defense ideas in Sun Tzu and Jomini, see Stephen Duane Biddle, "The Determinants of Offensiveness and Defensiveness in Conventional Land Warfare" (Ph.D. diss., Harvard University, 1992), 22–30. On Rousseau, see Stanley Hoffmann, *The State of War* (New York: Frederick A. Praeger, 1965), 74–82; and Robert E. Osgood and Robert C. Tucker, *Force, Order, and Justice* (Baltimore: Johns Hopkins University Press, 1967), 12.

16. J. F. C. Fuller, "What Is an Aggressive Weapon?" *English Review* 54 (June 1932): 601–5; B. H. Liddell Hart, "Aggression and the Problem of Weapons," *English Review* 55 (July 1932): 71–78, especially 72–73; B. H. Liddell Hart, *The Memoirs of Captain Liddell Hart* (London: Cassell & Co., 1965), 186; Marion William Boggs, *Attempts to Define and Limit "Aggressive" Armament in Diplomacy and Strategy* (Columbia: University of Missouri Studies, vol. 16, no. 1, 1941); Quincy Wright, *A Study of War* (Chicago: University of Chicago Press, 1965), 807–10, 1518–21 (the first edition of this book was published in 1942); Hoag, "On Stability in Deterrent Races"; and Schelling, *Arms and Influence,* 224–25, 234–35.

17. Jervis provided the first rigorous and systematic analysis of the international political consequences of technological change ("Cooperation under the Security Dilemma"), but the idea that war is more likely when offense enjoys a relative military advantage is the central thesis of Quester's 1977 book, *Offense and Defense in the International System.* Quester writes that "likelihoods of war are thus clearly influenced by how effective the offensive weapon seems to be, as compared with the defensive, and by how much the rival nations invest in each." He concludes that "offenses produce war and/or empire; defenses support independence and peace." *Ibid.,* 7, 208. Stephen Van Evera notes that he, Jack Snyder, and Shai Feldman independently developed elements of offense-defense theory during the late 1970s, although their work appeared later. Van Evera, *Causes of War,* 119n. See Snyder, "Civil-Military Relations and the Cult of the Offensive"; Shai Feldman, *Israeli Nuclear Deterrence: A Strategy for the 1980s* (New York:

Columbia University Press, 1982), 45–49; and Stephen Van Evera, "The Cult of the Offensive and the Origins of the First World War." Charles Glaser's application of offense-defense theory to nuclear policy appeared a few years later in Charles L. Glaser, *Analyzing Strategic Nuclear Policy* (Princeton: Princeton University Press, 1990).

18. Jervis, "Cooperation under the Security Dilemma," 169. The concept of the security dilemma is present, but less developed, in John H. Herz, "Idealist Internationalism and the Security Dilemma," *World Politics* 2 (January 1950): 157–80; and Herbert Butterfield, *History and Human Relations* (London: Collins, 1951). Charles Glaser reviews, clarifies, and extends the logic of Jervis's article in Glaser, "The Security Dilemma Revisited," 171–201.

19. Charles Glaser identifies at least three ways in which making one's adversary less secure can be self-defeating: first, by reducing the state's own military capability after arms competition results in new weapons deployments and larger forces; second, by increasing the value the adversary places on expansion, which makes it harder to deter; and, third, by wasting money. Glaser, "Security Dilemma Revisited," 174–83.

20. See Robert Jervis, *Perception and Misperception in International Politics* (Princeton: Princeton University Press, 1976), 62–76.

21. The realist tradition, Stephen Walt writes, "identifies and explains the central *problematique* in the field of international relations and sheds considerable light on a diverse array of important international phenomena. As a result, no serious scholar can safely disregard its arguments and implications. Even prominent critics of realist theory acknowledge its central place in the discipline, and many of its arguments are echoed by scholars who are not normally regarded as 'realists.' " Stephen M. Walt, "The Enduring Relevance of the Realist Tradition," in *Political Science: The State of the Discipline*, ed. Ira Katznelson and Helen V. Milner (New York: W. W. Norton, 2002), 197–234, at 198.

22. The labels "offensive" and "defensive" realism do not derive from offense-defense theory per se. The potential confusion is unavoidable because these terms have gained greater currency over alternative labels such as "aggressive" and "contingent" realism. On the debate between offensive and defensive realism and other relevant debates within realism, see Sean M. Lynn-Jones and Steven E. Miller, "Preface," in *The Perils of Anarchy: Contemporary Realism and International Security*, ed. Michael E. Brown, Sean M. Lynn-Jones, and Steven E. Miller (Cambridge: MIT Press, 1995), ix–xiii.; Benjamin Frankel, "Restating the Realist Case: An Introduction," in *Realism: Restatements and Renewal*, ed. Benjamin Frankel (London: Frank Cass, 1996), xv–xx; Colin Elman, "Horses for Courses: Why *Not* Neorealist Theories of Foreign Policy?" *Security Studies* 6 (Autumn 1996): 7–53, at 21–32; Stephen G. Brooks, "Dueling Realisms," *International Organization* 51 (Summer 1997): 445–77; Gideon Rose, "Neoclassical Realism and Theories of Foreign Policy," *World Politics* 51 (October 1998): 144–72; Sean M. Lynn-Jones, "Realism and America's Rise: A Review Essay," *International Security* 23 (Fall 1998): 157–82; Jeffrey W. Taliaferro, "Security Seeking under Anarchy: Defensive Realism Revisited," *International Security* 25 (Winter 2000/2001): 128–61; and Stephen M. Walt, "Enduring Relevance of the Realist Tradition." On the relationship between offense-defense theory and realism, see Van Evera, *Causes of War,* 7–11, 117, 255–56; Glaser and Kaufmann, "What Is the Offense-Defense Balance?" 48–49; Lynn-Jones, "Offense-Defense Theory and Its Critics," 660n, 664–65; and Glaser, "Realists as Optimists," 54, 60–64.

23. The case for defensive realism is found in Glaser, "Realists as Optimists." Also see Jervis, "Cooperation under the Security Dilemma"; Van Evera, *Causes of War;* Jack Sny-

der, *Myths of Empire: Domestic Politics and International Ambition* (Ithaca: Cornell University Press, 1991); Kydd, "Sheep in Sheep's Clothing"; Lynn-Jones, "Realism and America's Rise"; and Taliaferro, "Security Seeking under Anarchy."

24. The following discussion draws primarily on Jervis, "Cooperation under the Security Dilemma," 187–99; and Glaser, "Security Dilemma Revisited," 185–87.

25. This echoes George Quester, who writes, "If both sides are primed to reap advantages by pushing into each other's territory, war may be extremely likely whenever political crisis erupts. If the defense holds the advantage, by contrast, each side in a crisis will probably wait a little longer, in hopes that the others will foolishly take the offensive." Quester, *Offense and Defense in the International System*, 7. On the dangers of first-strike advantages, see Thomas C. Schelling and Morton H. Halperin, *Strategy and Arms Control* (New York: Twentieth Century Fund, 1961), 14–16; and Thomas C. Schelling, *The Strategy of Conflict* (New York: Oxford University Press, 1963), chap. 9.

26. The offense-defense balance has been widely employed in explanations of alliance behavior. George Quester argues that "hardening" and "congealing" of alliances "must be traced to our continuing villain, offensive military technology." See Quester, *Offense and Defense in the International System*, 105–6. Stephen Walt argues that offensive advantages make offensive capabilities more threatening and thus are more likely to provoke a balancing alliance. Walt adds, "Alliance formation becomes more frenetic when the offense is believed to have the advantage: great powers will balance more vigorously, and weak states will bandwagon more frequently. A world of tight alliances and few neutral states is the likely result." See Walt, *Origins of Alliances*, 24–25n, and 165–67. Also arguing that offensive advantages lead to tighter alliances are Thomas J. Christensen and Jack Snyder, "Chain Gangs and Passed Bucks," 137–68; and Thomas J. Christensen, "Perceptions and Alliances in Europe." Jervis suggests that polarized alliances make war more likely, but he does not develop this argument. Jervis, "Cooperation under the Security Dilemma," 189.

27. Jervis, "Cooperation under the Security Dilemma," 199–201. See also Glaser, "When Are Arms Races Dangerous?" 56.

28. The case for offensive realism is Mearsheimer, *Tragedy of Great Power Politics*. Other squarely offensive realist works include Fareed Zakaria, *From Wealth to Power: The Unusual Origins of America's World Role* (Princeton: Princeton University Press, 1998); and Eric J. Labs, "Beyond Victory: Offensive Realism and the Expansion of War Aims," *Security Studies* 6 (Summer 1997): 1–49. Elements of offensive realism are found in Gilpin, *War and Change in World Politics;* Randall L. Schweller, "Bandwagoning for Profit: Bringing the Revisionist State Back In," *International Security* 19 (Summer 1994): 72–107; and Schweller, "Neorealism's Status-Quo Bias."

29. On microfoundations, James Morrow writes, "Systemic theories attempt to explain international conflict as a consequence of the properties of the international system. Microfoundations address the actor-level processes that underlie such systemic theories." James D. Morrow, "International Conflict: Assessing the Democratic Peace and Offense-Defense Theory," in Katznelson and Milner, *Political Science*, 172–96, at 173.

30. James D. Fearon, "Rationalist Explanations for War," *International Organization* 49 (Summer 1995): 379–414, at 401–4.

31. Alexander Wendt, *Social Theory of International Politics* (New York: Cambridge University Press, 1999), 357–63. Wendt writes: "When defensive technology has a signifi-

cant (and know) advantage, or when offensive technology is dominant but unuseable, as with nuclear weapons under Mutual Assured Destruction, then states are constrained from going to war and thus, ironically, may be willing to trust each other enough to take on a collective identity" (358). Others scholars have suggested that the modified realist balance-of-threat account of international alliance behavior, which incorporates the offense-defense balance, is a fundamentally constructivist explanation: if what constitutes a "threat" is subjectively determined between states, then the implications of an offense-defense balance may also be socially constructed—and potentially malleable. See Fearon, "The Offense-Defense Balance and War Since 1648," 402–3; Ted Hopf, "The Promise of Constructivism in International Relations Theory," *International Security* 23 (Summer 1998): 171–200, at 186–88; Kier, "Culture and Military Doctrine"; and Alexander Wendt, "Anarchy Is What States Make of It: The Social Construction of Power Politics," *International Organization* 46 (Spring 1992): 391–425.

32. Van Evera, *Causes of War,* 119. The author argues, however, that offense-defense theory remains underappreciated. Ibid., 117.

33. See Boggs, *Attempts to Define and Limit "Aggressive" Armament.* Two influential commentators on military affairs at the time, J. F. C. Fuller and B. H. Liddell Hart, debated the offense-defense impact of specific weapons and technologies. See Fuller, "What Is an Aggressive Weapon?"; Liddell Hart, "Aggression and the Problem of Weapons," 71–78, especially 72–73; and Liddell Hart, *Memoirs of Captain Liddell Hart,* 186.

34. Merze Tate, *The United States and Armaments* (Cambridge: Harvard University Press, 1948), 108.

35. Wright, *Study of War,* 807. Also see ibid., 808–10, 1518–21. The original text is from 1942.

36. Schelling and Halperin, *Strategy and Arms Control.*

37. Jonathan Dean, "Alternative Defense: An Answer to NATO's Central Front Problems?" *International Affairs* 64 (Winter 1987–88): 61–82; Stephen J. Flanagan, "Nonprovocative and Civilian-Based Defenses," in *Fateful Visions: Avoiding Nuclear Catastrophe,* ed. Joseph S. Jr. Nye, Graham T. Allison, and Albert Carnesale (Cambridge, Mass.: Ballinger, 1988), 93–109; Anders Boserup and Robert Neild, *The Foundations of Defensive Defense* (New York: St. Martin's Press, 1990); and Bjørn Møller, *Common Security and Nonoffensive Defense: A Neorealist Perspective* (Boulder, Colo.: Lynne Rienner, 1992).

38. See Project on Defense Alternatives, "Confidence Building Defense," http://www.comw.org/pda/confblg.html; and Geoffrey Wiseman, *Concepts of Non-Provocative Defence: Ideas and Practices in International Security* (New York: Palgrave, 2002).

39. Edward J. Laurance, *Light Weapons and Intrastate Conflict: Early Warning Factors and Preventive Action* (New York: Carnegie Corp., 1998); Jeffrey Boutwell and Michael T. Klare, *Light Weapons and Civil Conflict: Controlling the Tools of Violence* (Lanham, Md.: Rowman & Littlefield, 1999).

40. Paul Bracken, "America's Maginot Line," *Atlantic Monthly,* December 1998, 85–93, at 87.

41. Charles L. Glaser and Steve Fetter, "National Missile Defense and the Future of U.S. Nuclear Weapons Policy," *International Security* 26 (Summer 2001): 40–92.

42. See Scott D. Sagan and Kenneth N. Waltz, *The Spread of Nuclear Weapons: A Debate Renewed* (New York: W. W. Norton, 2003).

43. See Jervis, "Cooperation under the Security Dilemma," 199–201; Glaser and Kaufmann, "What Is the Offense-Defense Balance?" 44; and Van Evera, "Offense, De-

fense, and the Causes of War," 40. Other proponents of offense-defense theory are less sanguine about the capacity of arms control to reduce the likelihood of war, pointing out that arms control may be possible only when it is not necessary. I thank Sean Lynn-Jones for comments on this point.

44. See, for example, Eliot A. Cohen, "A Revolution in Warfare," *Foreign Affairs* 75 (March/April 1996): 37–54, at 45; "Select Enemy. Delete," *Economist,* March 8, 1997, 21–24, at 21; James R. Blaker, *Understanding the Revolution in Military Affairs: A Guide to America's 21st Century Defense* (Washington, D.C.: Progressive Policy Institute, 1997), 15.

45. Current debates about whether East Asia is primed for conflict also revolve in part around competing interpretations of the offense-defense balance of technology. For example, some argue that the security dilemma and the offense-defense balance predict spirals of tension between China and Japan. Others predict stability because of the region's high defense dominance. See Thomas J. Christensen, "China, the U.S.-Japan Alliance, and the Security Dilemma in East Asia," *International Security* 23 (Spring 1999): 49–80; and Jennifer M. Lind and Thomas J. Christensen, "Correspondence: Spirals, Security, and Stability in East Asia," *International Security* 24 (Spring 2000): 190–200.

46. Richard K. Betts, "The Soft Underbelly of American Primacy: Tactical Advantages of Terror," *Political Science Quarterly* 117 (Spring 2002): 19–36. Betts warns, however, that the benefits of counteroffensive operations against terrorists need to be weighed against the costs of alienating civilian populations through collateral damage. Ibid., 33.

47. Jack S. Levy, "The Offensive/Defensive Balance of Military Technology: A Theoretical and Historical Analysis," *International Studies Quarterly* 28 (June 1984): 219–38, at 234.

48. Van Evera, *Causes of War,* chap. 6. Van Evera's tests find strong support for offense-defense theory.

49. See Glaser, "Realists as Optimists"; Lynn-Jones, "Offense-Defense Theory and Its Critics"; and Glaser and Kaufmann, "What Is the Offense-Defense Balance?" The contributions of these works are discussed in greater detail in the next chapter.

50. Betts, "Must War Find a Way?"

51. Lynn-Jones, "Offense-Defense Theory and Its Critics," 691; Glaser, "The Security Dilemma Revisited," 200.

52. Adams, "Attack and Conquer?"; Gortzak, Haftel, and Sweeney, "Offense-Defense Theory."

53. Van Evera, *Causes of War,* chap. 6; Stephen Biddle, *Military Power: Explaining Victory and Defeat in Modern Battle* (Princeton: Princeton University Press, 2004); Biddle, "Rebuilding the Foundations of Offense-Defense Theory." Other recent empirical evaluations of the theory include Liberman, "Offense-Defense Balance"; Lieber, "Grasping the Technological Peace"; and Russell, "Persian Gulf Proving Grounds."

54. Stephen Van Evera discusses the attributes of a good theory and why offense-defense theory has those attributes in *Causes of War,* 3–4, 190–92; and Van Evera, *Guide to Methods for Students of Political Science* (Ithaca: Cornell University Press, 1997), 17–21.

55. Among the many works on the case study method, see Alexander L. George and Andrew Bennett, *Case Studies and Theory Development in the Social Sciences* (Cambridge: MIT Press, forthcoming); Harry Eckstein, "Case Study and Theory in Political Science," in *Handbook of Political Science,* vol. 7, ed. Fred Greenstein and Nelson Polsby (Reading, Mass.: Addison-Wesley, 1975), 79–138; Alexander L. George and Timothy J. McKeown,

"Case Studies and Theories of Organizational Decision Making," in *Advances in Information Processing in Organizations* (Greenwich, Conn.: JAI Press, 1985); Robert K. Yin, *Case Study Research: Design and Methods,* 2d ed. (Thousand Oaks, Calif.: Sage, 1994); Van Evera, *Guide to Methods,* 49–88, 129–31; and John Gerring, "What Is a Case Study and What Is It Good for?" *American Political Science Review* 98 (May 2004): 341–54.

56. Extolling the essential complementarity of alternative research methods in the social sciences while seeking to correct the traditional prejudice against the case study method are Henry E. Brady and David Collier, eds., *Rethinking Social Inquiry: Diverse Tools, Shared Standards* (Lanham, Md.: Rowman & Littlefield, 2004); and George and Bennett, *Case Studies and Theory Development.* On the traditional disregard for case study method, see Yin, *Case Study Research,* 9–13. Although cast as an effort to develop a shared framework for social inquiry for both quantitative and qualitative analysis, a major earlier contribution overlooked many of the basic limitations of quantitative analysis and failed to appreciate the distinctive strengths of case study methods; see, Gary King, Robert O. Keohane, and Sidney Verba, *Designing Social Inquiry: Scientific Inference in Qualitative Research* (Princeton: Princeton University Press, 1994). Specifically critiquing *Designing Social Inquiry* are Timothy J. McKeown, "Case Studies and the Statistical Worldview: Review of King, Keohane, and Verba's *Designing Social Inquiry: Scientific Inference in Qualitative Research,*" *International Organization* 53 (Winter 1999): 161–90; Brady and Collier, *Rethinking Social Inquiry;* and George and Bennett, *Case Studies and Theory Development,* chap. 1.

57. See Ronald Rogowski, "The Role of Theory and Anomaly in Social-Scientific Inference," *American Political Science Review* 89 (June 1995): 467–70; and Dietrich Rueschemeyer, "Can One or a Few Cases Yield Theoretical Gains?" in *Comparative Historical Analysis in the Social Sciences,* ed. James Mahoney and Dietrich Rueschemeyer (New York: Cambridge University Press, 2003), 305–36. Gerring agrees that one of the most important comparative strengths of case studies is that they are generally more useful when useful variance is available for only a single case or a small number of cases. Gerring, "What Is a Case Study?" 350–52.

58. On the comparative advantages of case studies vis-à-vis statistical methods and formal models see George and Bennett, *Case Studies and Theory Development,* chap. 1.

59. On drawing the implications of case findings for theory, see George and Bennett, *Case Studies and Theory Development,* chap. 6.

60. On congruence procedure see George and Bennett, *Case Studies and Theory Development,* chap. 9; Van Evera, *Guide to Methods,* 58–63; and George and McKeown, "Case Studies and Theories," 29–34.

61. On process tracing see George and Bennett, *Case Studies and Theory Development,* chap. 10; Van Evera, *Guide to Methods,* 64–67; George and McKeown, "Case Studies and Theories," 34–41; and Andrew Bennett and Alexander L. George, "Case Studies and Process Tracing in History and Political Science: Similar Strokes for Different Foci," in *Bridges and Boundaries: Historians, Political Scientists, and the Study of International Relations,* ed. Colin Elman and Miriam Fendius Elman (Cambridge: MIT Press, 2001), 137–66, especially 144–53.

62. King, Keohane, and Verba view process tracing as simply a means of increasing the number of theoretically relevant observations. By breaking down a sequence of events or causal processes into multiple independent and dependent variables, process tracing "can help to overcome the dilemmas of small-*n* research and enable investigators and their readers to increase their confidence in the findings of social science."

King, Keohane, and Verba, *Designing Social Inquiry,* 227. George and Bennett disagree with this characterization of process tracing as little more than a method of creating a larger number of data points: "In fact, process tracing is fundamentally different from statistical analysis because it focuses on sequential processes within a particular histori-cal case, not on correlations of data across cases. This has important implications for theory testing: a single unexpected piece of process tracing evidence can require alter-ing the historical interpretation and theoretical significance of a case, whereas several such cases may not greatly alter the findings concerning statistical estimates of param-eters for a large population." George and Bennett, *Case Studies and Theory Development,* chap. 1. See also Brady and Collier's distinction between data-set observations and causal-process observations. Brady and Collier, *Rethinking Social Inquiry,* 252–55.

63. Christopher H. Achen and Duncan Snidal, "Rational Deterrence Theory and Comparative Case Studies," *World Politics* 41 (January 1989): 160–61; Barbara Geddes, "How the Cases You Choose Affect the Answers You Get: Selection Bias in Comparative Politics," in *Political Analysis,* vol. 2, ed. James A. Stimson (Ann Arbor: University of Michigan Press, 1990), 131–50; and King, Keohane, and Verba, *Designing Social Inquiry,* 128–39. For critical responses to these warnings see David Collier, James Mahoney, and Jason Seawright, "Claiming Too Much: Warnings about Selection Bias," in Brady and Collier, *Rethinking Social Inquiry,* 85–102; David Collier, "Translating Quantitative Meth-ods for Qualitative Researchers: The Case of Selection Bias," *American Political Science Review* 89 (June 1995); and David Collier and James Mahoney, "Insights and Pitfalls: Se-lection Bias in Qualitative Research," *World Politics* 49 (October 1996): 56–91.

64. Arguing that the methodological implications of selecting on the dependent variable are less problematic for case study work than has often been claimed is Dou-glas Dion, "Evidence and Inference in the Comparative Case Study," *Comparative Politics* 30 (January 1998): 127–45. See also the critical responses cited in note 63.

65. This method of "critical-case" analysis is developed in Eckstein, "Case Study and Theory" and refined and extended in Van Evera, *Guide to Methods,* 30–34, and George and Bennett, *Case Studies and Theory Development,* chap. 6. As a practical matter, each case differs in its degree of "toughness" for offense-defense theory, as that determina-tion depends not only on the likelihood of the theory's predictions for the given case (the "certainty" of its predictions) but also on other theories' predictions for the case (the "uniqueness" of its predictions). The strongest tests of predictions are those that are highly certain and highly unique, as supporting evidence would strongly corrobo-rate the theory and disconfirming evidence would badly damage it. Van Evera, *Guide to Methods,* 31.

66. An important additional question—Do leaders even think in offense-defense terms?—is implicitly addressed within the case chapters.

67. Van Evera, "Offense, Defense, and the Causes of War," 197.

Chapter 1. The Offense-Defense Balance

1. Stephen Van Evera includes in his book a section on "Qualifications: When Of-fensive Doctrines and Capabilities Cause Peace," but these exceptions apply only to sit-uations where a status quo power possessing offensive capabilities faces an aggressor state. He writes, "Symmetrical offense dominance—a situation where both sides have

strong offensive capabilities—is always more dangerous than symmetrical defense dominance, other things being equal." Stephen Van Evera, *Causes of War: Power and the Roots of Conflict* (Ithaca: Cornell University Press, 1999), 152.

2. See Sean M. Lynn-Jones, "Realism, Security, and Offense-Defense Theories: The Implications of Alternative Definitions of the Offense-Defense Balance" (paper presented at the annual meeting of the American Political Science Association, Boston, September 3–6, 1998).

3. For slightly different definitions see Edward Luttwak and Stuart L. Koehl, *The Dictionary of Modern War* (New York: Gramercy, 1998), 568, 442, 598.

4. On the problem of integrating across levels of warfare see Jonathan Shimshoni, "Technology, Military Advantage and World War I: A Case for Military Entrepreneurship," *International Security* 15 (Winter 1990–91): 191–93. Proponents who acknowledge the difficulty of integrating across levels of warfare respond that lower-level changes will usually shift the balance in the same direction at all higher levels, that the magnitude of the effect will "depend on . . . complex combinations of operational and tactical constraints and opportunities" that can be evaluated with detailed net assessments, and that the overall offense-defense balance can usually be understood by looking at key battles or campaigns. Charles L. Glaser and Chaim Kaufmann, "What Is the Offense-Defense Balance and Can We Measure It?" *International Security* 22 (Spring 1998): 44–82, at 73–74.

5. Agreeing with the relevance of the operational level, although for somewhat different reasons, are Stephen Biddle, "Rebuilding the Foundations of Offense-Defense Theory," *Journal of Politics* 63 (August 2001): 741–74, at 747–48; and Karen Ruth Adams, "Attack and Conquer? International Anarchy and the Offense-Defense-Deterrence Balance," *International Security* 28 (Winter 2003–4): 45–83, at 50–51.

6. The logical and methodological problems of previous attempts to define the balance were first highlighted in Jack S. Levy, "The Offensive/Defensive Balance of Military Technology: A Theoretical and Historical Analysis," *International Studies Quarterly* 28 (June 1984): 222–30.

7. For variations on this definition, see Robert Jervis, "Cooperation under the Security Dilemma," *World Politics* 30 (January 1978): 167–214, at 188; Charles L. Glaser, "Realists as Optimists: Cooperation as Self-Help," *International Security* 19 (Winter 1994–95): 50–90, at 61; Sean M. Lynn-Jones, "Offense-Defense Theory and Its Critics," *Security Studies* 4 (Summer 1995): 660–91, at 665; Glaser and Kaufmann, "What Is the Offense-Defense Balance?" 3, 7–10; and Robert Gilpin, *War and Change in World Politics* (Cambridge: Cambridge University Press, 1981), 62.

8. Biddle, "Rebuilding the Foundations of Offense-Defense Theory," 749n (emphasis in original). Biddle proposes defining the balance in terms of more observable measures, such as attacker casualties per defender casualty or attacker casualties per square kilometer of territory conquered. Ibid.

9. Stephen Van Evera partially dissents from the optimality assumption, arguing (as noted below) that force posture, doctrine, and strategy shape the offense-defense balance itself. However, this seems to undermine one of the potential strengths of offense-defense theory: its ability to explain patterns of state behavior based on structural incentives and constraints. See Van Evera, *Causes of War,* 160–63; Glaser and Kaufmann, "What Is the Offense-Defense Balance?" 55–57. For a critique of the optimality assumption and a reply by proponents, see James W. Davis and others, "Correspondence: Tak-

ing Offense at Offense-Defense Theory," *International Security* 23 (Winter 1998–99): 179–206, at 192–94, 200–202.

10. Adopting the core approach are George Quester, Robert Jervis, and Sean Lynn-Jones. Although Jervis cites technology and geography as the two main factors that determine whether offense or defense has the advantage, technology appears more significant for understanding the severity of the security dilemma. See Jervis, "Cooperation under the Security Dilemma," 194–96.

11. For a similar formulation see Lynn-Jones, "Offense-Defense Theory and Its Critics," 667.

12. Jervis, "Cooperation under the Security Dilemma," 183–85, 194–96; Van Evera, *Causes of War*, 163; and Glaser and Kaufmann, "What Is the Offense-Defense Balance?" 64–66.

13. Ted Hopf, "Polarity, the Offense-Defense Balance, and War," *American Political Science Review* 85 (June 1991): 477–78; Lynn-Jones, "Offense-Defense Theory and Its Critics," 669; and Glaser and Kaufmann, "What Is the Offense-Defense Balance?" 67–68.

14. Glaser and Kaufmann, "What Is the Offense-Defense Balance?" 66–67, 67n68. Jervis refers to nationalism as a "quasi-geographical aid to the defense." Jervis, "Cooperation under the Security Dilemma," 195.

15. Van Evera, *Causes of War*, 163–64.

16. Ibid., 164–66; and Hopf, "Polarity, the Offense-Defense Balance, and War," 477–78. Hopf refers to *beliefs* about balancing or bandwagoning as a factor affecting the offense-defense balance.

17. See Glaser and Kaufmann, "What Is the Offense-Defense Balance?" 68–70. See also James D. Fearon, "The Offense-Defense Balance and War since 1648" (paper prepared for annual meeting of the International Studies Association, Chicago February 21–25, 1995), 11–12; and Keir Lieber and David Edelstein, "Offense-Defense Theory and Alliance Behavior" (paper prepared for presentation at the Revolution in Military Affairs Conference, Naval Postgraduate School, Monterey, Calif., August 26–29, 1996), 23–24.

18. Van Evera, *Causes of War*, 161n162.

19. Glaser and Kaufmann, "What Is the Offense-Defense Balance?" 66.

20. Van Evera, *Causes of War*, 162. Van Evera also argues that wartime military operations can change the offense-defense balance. For example, "Aggressive operations can corrode key enemy defenses, and reckless operations can expose one's own defenses." Ibid.

21. Glaser and Kaufmann, "What Is the Offense-Defense Balance?" 41.

22. See Lynn-Jones, "Offense-Defense Theory and Its Critics," 668.

23. Additional measurement problems arise if the balance depends on specifying the costs of fighting the attacker is willing to incur, the value of the territory for the attacker, and the amount of territory the attacker is trying to take. For an argument that these specifications are required, see Glaser and Kaufmann, "What Is the Offense-Defense Balance?" 51–54. Logically this entails not only a different balance for every state depending on whether it is the attacker or the defender, but also an infinite number of balances depending on the specification of national costs, values, and goals!

24. For example, Van Evera's composite measure of the balance for European great powers since 1789 is based on his own "author's estimates," which are presented with-

out much explanation and are impossible to replicate. Van Evera, *Causes of War,* 168–92.

25. See Levy, "Offensive/Defensive Balance," 219–38; John J. Mearsheimer, *Conventional Deterrence* (Ithaca: Cornell University Press, 1983), 25–27; Samuel P. Huntington, "U.S. Defense Strategy: The Strategic Innovations of the Reagan Years," in *American Defense Annual, 1987–1988,* ed. Joseph Kruzel (Lexington, Mass.: Lexington Books, 1987), 35–37; Shimshoni, "Technology, Military Advantage and World War I," 190–91; and Colin S. Gray, *Weapons Don't Make War: Policy, Strategy, and Military Technology* (Lawrence: University Press of Kansas, 1993), chap. 2.

26. Lynn-Jones, "Offense-Defense Theory and Its Critics," 675.

27. Sean M. Lynn-Jones, "Preface," in *Offense, Defense, and War,* ed. Michael E. Brown and others (Cambridge: MIT Press, 2004), xi–xxxviii, at xiv. Elsewhere, Lynn-Jones writes, "Even if one does focus on particular weapons systems, some types may make offensive action easier and less costly than others." Lynn-Jones, "Offense-Defense Theory and Its Critics," 676. Glaser and Kaufmann write, "In a forthcoming article we argue that offensive and defensive weapons and force postures are generally distinguishable." "What Is the Offense-Defense Balance?" 80n.

28. See Huntington, "U.S. Defense Strategy," 36; and Albert Wohlstetter, "The Political and Military Aims of Offense and Defense Innovation," in *Swords and Shields,* ed. Fred S. Hoffman, Albert Wohlstetter, and David S. Yost (Lexington, Mass.: D.C. Heath, 1987), 4; David Goldfischer, *The Best Defense: Policy Alternatives for U.S. Nuclear Security from the 1950s to the 1990s* (Ithaca: Cornell University Press, 1993), chap. 1; Marion William Boggs, *Attempts to Define and Limit "Aggressive" Armament in Diplomacy and Strategy* (Columbia: University of Missouri Studies, vol. 16, no. 1, 1941); Lynn-Jones, "Offense-Defense Theory and Its Critics," 676–77.

29. Boggs, *Attempts to Define and Limit "Aggressive" Armament,* 46.

30. Ibid., 47–48.

31. Ibid., 49.

32. Ibid., 56, 58n.

33. Ibid., 84–85.

34. J. F. C. Fuller, "What Is an Aggressive Weapon?" *English Review* (June 1932): 601–5, at 601; B. H. Liddell Hart, "Aggression and the Problem of Weapons," *English Review* 55 (July 1932): 71–78, especially 72–73; B. H. Liddell Hart, *The Memoirs of Captain Liddell Hart* (London: Cassell & Co., 1965), 186.

35. J. F. C. Fuller, "Aggression and Aggressive Weapons: The Absurdity of Qualitative Disarmament," *Army Ordnance* 14 (1933): 7–11, at 9; Brian Holden Reid, *J. F. C. Fuller: Military Thinker* (New York: St. Martin's Press, 1987), 67.

36. Reid, *J. F. C. Fuller,* 57 and chap. 7; J. F. C. Fuller, *The Reformation of War* (New York: E. P. Dutton, 1923), 152–69; John J. Mearsheimer, *Liddell Hart and the Weight of History* (Ithaca: Cornell University Press, 1988), 33–35.

37. Mearsheimer, *Liddell Hart,* 36, and 35n; Liddell Hart, *Memoirs,* 186. Mearsheimer argues, however, that Liddell Hart's military views were increasingly warped by his policy preferences in the years leading up to World War II. Mearsheimer, *Liddell Hart,* 6–7, 110–23.

38. Cited in Mearsheimer, *Liddell Hart,* 114.

39. George H. Quester, *Offense and Defense in the International System* (New York: John Wiley, 1977), 2.

40. Ibid., 3.

41. Ibid., 28–29, 30, 34 versus 31, 34. In addition to mobility, Quester cites the temporary potency of a weapon as a technological characteristic favoring offense. "The ability of man to fly (temporarily), or to be totally mobilized into armed forces (temporarily), thus may favor the offense and threaten peace." However, Quester does not explain how technology relates to temporary potency. Ibid., 4.

42. Jervis, "Cooperation under the Security Dilemma," 197.

43. Ibid., 198.

44. Ibid., 201–2.

45. Ibid., 203.

46. Lynn-Jones, "Offense-Defense Theory and Its Critics," 674–75.

47. Ibid., 675–76.

48. Ibid., 667, 676.

49. Van Evera, *Causes of War,* 160.

50. Ibid., 160–61.

51. Glaser and Kaufmann discuss the offense-defense impact of six major areas of technology: mobility, firepower, protection, logistics, communication, and detection. Their views on mobility are discussed here, those on firepower in the next section. Glaser and Kaufmann are unable to judge the general offense-defense impact of the remaining four areas of technology: "The effects of innovations in protection, logistics, communication, and detection are more varied, depending on how specific innovations interact with force behavior; those whose full benefit can be realized only by non-advancing forces or only against advancing ones will favor defense, whereas those with benefits that are equally available to both advancing and nonadvancing forces will favor offense (at least compared with technologies of unequal usefulness)." For example, the authors state that landline telephones favored defense while portable radios favored offense, and early radar was more favorable to defense than is modern radar. Glaser and Kaufmann, "What Is the Offense-Defense Balance?" 64.

52. Ibid., 61. See also 79–80.

53. Ibid., 61.

54. Ibid., 62–63.

55. Ibid., 62–63.

56. See Mearsheimer, *Conventional Deterrence,* 25–26. Glaser and Kaufmann acknowledge that the impact of mobility is indeterminate with regard to the exploitation stage of an offensive. "What Is the Offense-Defense Balance?" 63.

57. See Stephen Biddle, "The Determinants of Offensiveness and Defensiveness in Conventional Land Warfare" (Ph.D. diss., Harvard University, 1992), 68–77; and Stephen Biddle, *Military Power: Explaining Victory and Defeat in Modern Battle* (Princeton: Princeton University Press, 2004), 35–44.

58. The mobility-firepower distinction is also used in the arms control policy community. See, for example, Catherine M. Kelleher, "Indicators of Defensive Intent in Conventional Force Structures and Operations in Europe," in *Military Power in Europe: Essays in Memory of Jonathan Alford,* ed. Lawrence Freedman (New York: St. Martin's Press, 1990), 159–78; Anders Boserup, "Mutual Defensive Superiority and the Problem of Mobility Along an Extended Front," in *The Foundations of Defensive Defense,* ed. Anders Boserup and Robert Neild (New York: St. Martin's Press, 1990), 63–78; and Gunilla Hesolf, "New Technology Favors the Defense," *Bulletin of the Atomic Scientists* 44, no. 7 (September 1988).

59. Boggs, *Attempts to Define and Limit "Aggressive" Armament,* 84.

60. Quincy Wright, *A Study of War* (Chicago: University of Chicago Press, 1965), 805–10.

61. Quester, *Offense and Defense,* 3–4, 15–17, 31–35, 63.

62. Ibid., 45–48, 60, 100–122.

63. Jervis, "Cooperation under the Security Dilemma," 203.

64. Ibid., 191–92, 197.

65. Van Evera, *Causes of War,* 160–61.

66. Glaser and Kaufmann, "What Is the Offense-Defense Balance?" 64.

67. The authors qualify their conclusions about the offense-defense impact of firepower, with exceptions for "when specific firepower innovations are differentially useful against defenders." Ibid., 64. Note, however, that this seems equivalent to claiming that firepower improvements favor defense except when they favor offense.

68. See Biddle, *Military Power,* 44–48.

Chapter 2. The Railroad Revolution

1. Some proponents of offense-defense theory view railroads as an exception to the mobility-favors-offense prediction. According to the argument, railroad mobility is unlike other kinds of mobility because retreating defenders can destroy elaborate rail networks more easily than advancing attackers can extend them. However, railroads should not be excluded as a test case for at least three reasons. First, a rigorous assessment of the impact of railroads on the offense-defense balance would be useful in itself because proponents are divided on the question. Second, an empirical finding that railroads favored defenders on the whole would considerably diminish our confidence in the basic mobility-favors-offense hypothesis. The fact of the matter is that railroads marked a revolutionary improvement in mobility. If only large changes in technology have recognizable effects on the offense-defense balance and international politics, then the relevant data set of such changes is already quite small. Excluding railroads from the universe of mobility-enhancing cases would at best undermine the explanatory relevance of the theory and could at worst be interpreted as merely an ad hoc step to address the theory's shortcomings. Third, we can evaluate offense-defense theory according to how perceptions of railroads affect political behavior, which remains relevant even if railroads are excluded from an evaluation of predicted military outcomes. See Charles L. Glaser and Chaim Kaufmann, "What Is the Offense-Defense Balance and Can We Measure It?" *International Security* 22 (Spring 1998): 44–82, at 63; George H. Quester, *Offense and Defense in the International System* (New York: John Wiley & Sons, 1977), chap. 8; Jack Snyder, "Civil-Military Relations and the Cult of the Offensive, 1914 and 1984," *International Security* 9 (1984): 108–46; and Stephen Van Evera, "Offense, Defense, and the Causes of War," *International Security* 22 (Spring 1998): 5–43, at 16–17.

2. Dennis E. Showalter, *Railroads and Rifles: Soldiers, Technology, and the Unification of Germany* (Hamden, Conn.: Archon, 1975), 37.

3. Edwin A. Pratt, *The Rise of Rail-Power in War and Conquest, 1833–1914* (Philadelphia: J. B. Lippincott, 1916), 8; Showalter, *Railroads and Rifles,* 37–38; Arden Bucholz, *Moltke, Schlieffen, and Prussian War Planning* (New York: Berg, 1991), 44–46; and John Merriman, *A History of Modern Europe: From the French Revolution to the Present* (New York: W. W. Norton, 1996), 741–42.

4. Pratt, *Rise of Rail-Power,* 9–13; John Westwood, *Railways at War* (San Diego: Howell-North, 1980), 14–16; Patrick Turnbull, *Solferino: The Birth of a Nation* (New York: St. Martin's Press, 1985), chaps. 7–9; R. Ernest Dupuy and Trevor N. Dupuy, *The Harper Encyclopedia of Military History: From 3500 B.C. to the Present* (New York: HarperCollins, 1993), 907–8; and Richard Brooks, "The Italian Campaign of 1859," *Military History* 16 (June 1999). The quotation is from Pratt, *Rise of Rail-Power,* 12.

5. James M. McPherson, *Battle Cry of Freedom: The Civil War Era* (New York: Ballantine Books, 1988), 12.

6. Russell F. Weigley, *A Great Civil War: A Military and Political History, 1861–1865* (Bloomington: Indiana University Press, 2000), 31, 35.

7. On railroads in the Civil War, Robert C. Black, III, *The Railroads of the Confederacy* (Chapel Hill: University of North Carolina Press, 1998); and George Edgar Turner, *Victory Rode the Rails: The Strategic Place of the Railroads in the Civil War* (Lincoln: University of Nebraska Press, 1992) remain essential. More recently, see John E. Clark, Jr., *Railroads in the Civil War: The Impact of Management on Victory and Defeat* (Baton Rouge: Louisiana State University Press, 2001); and Robert G. Angevine, *The Railroad and the State: War, Politics, and Technology in Nineteenth-Century America* (Stanford: Stanford University Press, 2004), chap. 7. Also important are Roger Pickenpaugh, *Rescue by Rail: Troop Transfer and the Civil War in the West, 1863* (Lincoln: University of Nebraska Press, 1998); Christopher R. Gabel, *Railroad Generalship: Foundations of Civil War Strategy* (Fort Leavenworth, Kans.: U.S. Army Command and General Staff College, 1997); Jeffrey N. Lash, *Destroyer of the Iron Horse: General Joseph E. Johnston and Confederate Rail Transport, 1861–1865* (Kent, Ohio: Kent State University Press, 1991); Roy Meredith and Arthur Meredith, *Mr. Lincoln's Military Railroads* (New York: W. W. Norton, 1979); George B. Abdil, *Civil War Railroads* (Seattle: Superior Publishing Co., 1961); Angus J. Johnston, II, "Virginia Railroads in April 1861," *Journal of Southern History* 23 (August 1957): 307–30; Thomas Weber, *The Northern Railroads in the Civil War, 1861–1865* (Bloomington: Indiana University Press, 1952); Charles W. Ramsdell, "The Confederate Government and the Railroads," *American Historical Review* 22 (July 1917): 794–810; Margaret E. Wagner, Gary W. Gallagher, and Paul Finkelman, eds., *The Library of Congress Civil War Desk Reference* (New York: Simon & Schuster, 2002), 349–53; and David S. Heidler and Jeanne T. Heidler, *Encyclopedia of the American Civil War: A Political, Social, and Military History* (New York: W. W. Norton, 2000), 1591–98.

8. See Black, *Railroads of the Confederacy,* chap. 1; Turner, *Victory Rode the Rails,* chap. 2; and Angevine, *Railroad and the State,* chap. 7. Also see Clark, *Railroads in the Civil War,* 18–21; Lash, *Destroyer of the Iron Horse,* 45–48; and McPherson, *Battle Cry of Freedom,* 91n25.

9. Ramsdell, "Confederate Government and the Railroads," 796–97; Black, *Railroads of the Confederacy,* 8–9. Black writes, "Everywhere through Dixie railroads were stretching iron fingers toward one another, but not yet everywhere had they joined hands" (9).

10. Heidler and Heidler, *Encyclopedia of the American Civil War,* 1591; Black, *Railroads of the Confederacy,* chaps. 2, 5–6, and 10; and Weber, *Northern Railroads in the Civil War,* chap. 2.

11. A thorough account of the Confederacy's deplorable management of the railroads is Clark, *Railroads in the Civil War,* especially chaps. 1 and 6. Also see Black, *Railroads of the Confederacy,* chaps. 6, 8, 9, 13, and 18; Lash, *Destroyer of the Iron Horse,* 182–86; and Ramsdell, "Confederate Government and the Railroads," 799–809.

12. Gabel, *Railroad Generalship*, 4–5; Turner, *Victory Rode the Rails*, 319–36; Weber, *Northern Railroads in the Civil War*, 199–204; Wagner, Gallagher, and Finkelman, *Library of Congress Civil War Desk Reference*, 351.

13. Angevine, *Railroad and the State*, 145.

14. William T. Sherman, *Memoirs of General W. T. Sherman* (New York: Charles L. Webster, 1891), 2:398–99.

15. Weber, *Northern Railroads in the Civil War*, 199–205; Turner, *Victory Rode the Rails*, 319–36; and Edward Hagerman, *The American Civil War and the Origins of Modern Warfare: Ideas, Organization, and Field Command* (Bloomington: Indiana University Press, 1988), 280–83.

16. Westwood, *Railways at War*, 17, 29; Gabel, *Railroad Generalship*; Clark, *Railroads in the Civil War*, 73.

17. Lash, *Destroyer of the Iron Horse*, 7–18; Turner, *Victory Rode the Rails*, 86–95; Black, *Railroads of the Confederacy*, 60–62.

18. Black, *Railroads of the Confederacy*, 137–47.

19. Clark, *Railroads in the Civil War*, 88–126; Black, *Railroads of the Confederacy*, 184–91; Gabel, *Railroad Generalship*, 5–6; Pickenpaugh, *Rescue by Rail*, 27–43.

20. Clark, *Railroads in the Civil War*, 141–212; Weber, *Northern Railroads in the Civil War*, 180–81; Turner, *Victory Rode the Rails*, 282–96.

21. Black, *Railroads of the Confederacy*, 180; Addington, *The Patterns of War since the Eighteenth Century* (Bloomington: Indiana University Press, 1994), 83–84.

22. See McPherson, *Battle Cry of Freedom*, 514–15; Heidler and Heidler, *Encyclopedia of the American Civil War*, 1597.

23. See Gabel, *Railroad Generalship*, 6–9; Black, *Railroads of the Confederacy*, 294; Pratt, *Rise of Rail-Power*, 15. Describing the impact of railroads in the Civil War Archer Jones writes, "Because retreating armies routinely disabled the railroads, the defender often had better strategic mobility than the invader." Archer Jones, *The Art of War in the Western World* (New York: Oxford University Press, 1987), 644. Edward Hagerman writes that the means to achieve a Southern victory "would be an exploitation of interior lines facilitated by the new technology of the railroad and the comparative advantage that it gave to the defense." Hagerman, *American Civil War*, 102.

24. Angevine, *Railroad and the State*, 142–43; Gabel, *Railroad Generalship*, 5; Westwood, *Railways at War*, 29.

25. See Arden Bucholz, *Moltke and the German Wars, 1864–1871* (New York: Palgrave, 2001), 77–102; and Showalter, *Railroads and Rifles*, 48–51.

26. Geoffrey Wawro, *The Austro-Prussian War: Austria's War with Prussia and Italy in 1866* (New York: Cambridge University Press, 1996), 50–57; Gordon A. Craig, *The Battle of Königgrätz: Prussia's Victory Over Austria, 1866* (Philadelphia: J. B. Lippincott, 1964); William Carr, *The Origins of the Wars of German Unification* (London: Longman, 1991), 136–38.

27. Wawro, *Austro-Prussian War*, 50–65.

28. Martin van Creveld, *Supplying War: Logistics From Wallenstein to Patton* (London: Cambridge University Press, 1977), 83–85; Thomas J. Adriance, *The Last Gaiter Button* (Westport, Conn.: Greenwood Press, 1987), 42–47; Showalter, *Railroads and Rifles*, 68–72; Pratt, *Rise of Rail-Power*, 104–5; and Westwood, *Railways at War*, 57.

29. Wawro, *Austro-Prussian War*, 75–77.

30. The two definitive accounts are Geoffrey Wawro, *The Franco-Prussian War: The German Conquest of France in 1870–1871* (New York: Cambridge University Press, 2003);

and Michael Howard, *The Franco-Prussian War: The German Invasion of France, 1870–1871*, 2d ed. (New York: Routledge, 2001). On the disastrous mobilization and concentration of the French Army in particular, see Adriance, *Last Gaiter Button;* and William Sherman, "French Mobilization in 1870," in *On the Road to Total War: The American Civil War and the German Wars of Unification, 1861–1871*, ed. Stig Förster and Jörg Nagler (Washington, D.C.: German Historical Institute, 1997), 283–94.

31. Wawro, *Franco-Prussian War,* 41, 47.

32. Ibid., 49–50.

33. On the important differences between the French and Prussian armies in 1870, especially the deficiencies of the former, see Howard, *Franco-Prussian War,* 1–39; Wawro, *Franco-Prussian War,* 41–64; Adriance, *Last Gaiter Button,* 19–38.

34. There is general consensus, but the most extensive argument is found in Adriance, *Last Gaiter Button.*

35. Ibid., 3–19, 119–37; Wawro, *Franco-Prussian War,* 65–229; Howard, *Franco-Prussian War,* 57–76, 120–223. The quotation is from Wawro, *Franco-Prussian War,* 48.

36. Van Creveld, *Supplying War,* 96–104; Westwood, *Railways at War,* 66.

37. Adriance, *Last Gaiter Button,* 47–54; Pratt, *Rise of Rail-Power,* 110–15.

38. David Stevenson, *Cataclysm: The First World War as Political Tragedy* (New York: Basic Books, 2004), 39–41; Hew Strachan, *The First World War: To Arms* (Oxford: Oxford University Press, 2001), 1:207; David Stevenson, *Armaments and the Coming of War: Europe, 1904–1914* (Oxford: Oxford University Press, 1996).

39. David Stevenson, "War by Timetable? The Railway Race Before 1914," *Past and Present* 162 (February 1999): 163–94; Strachan, *To Arms,* 194, 206; Stevenson, *Cataclysm,* 41–42; Van Creveld, *Supplying War,* 112.

40. The overall balance of power between Germany and its adversaries was not even, however. At the outbreak of war, the Triple Entente (France, Russia, and Britain) could put 182 divisions into the field against 136 divisions from Germany and Austria-Hungary. Hew Strachan, *The First World War* (New York: Viking, 2003), 46.

41. Stevenson, *Cataclysm,* 43.

42. Van Creveld, *Supplying War,* 126–35; Strachan, *To Arms,* 239–242; Stevenson, *Cataclysm,* 45.

43. Strachan, *To Arms,* 243; Van Creveld, *Supplying War,* 109–41; Archer Jones, *Art of War,* 434–38.

44. Strachan, *To Arms,* 297–98, 312; Stevenson, "War by Timetable?"

45. Stevenson, "War by Timetable?"

46. Strachan, *To Arms,* 297.

47. Norman Stone, *The Eastern Front, 1914–1917* (New York: Charles Scribner's Sons, 1975), chap. 3; Dennis E. Showalter, *Tannenberg: Clash of Empires* (Hamden, Conn.: Archon Books, 1991); Strachan, *To Arms,* 316–35; Jones, *Art of War,* 441–43.

48. Stephen Van Evera, *Causes of War: Power and the Roots of Conflict* (Ithaca: Cornell University Press, 1999), 185.

49. Pratt, *Rise of Rail-Power,* x; Van Creveld, *Supplying War,* 82–83.

50. E. G. Campbell, "Railroads in National Defense, 1829–1848," *Mississippi Valley Historical Review* 27 (December 1940): 361–78, at 362, 365, 368. See also James A. Ward, *Railroads and the Character of America, 1820–1887* (Knoxville: University of Tennessee Press, 1986), chap. 3.

51. Quoted in Campbell, "Railroads in National Defense," 372.

52. Quoted in Van Creveld, *Supplying War,* 88.

53. Showalter, *Railroads and Rifles*, 21–22.

54. Quoted in Pratt, *Rise of Rail-Power*, 7.

55. For additional evidence suggesting the dominant view of railroads as defensive in this early period, see Showalter, *Railroads and Rifles*, 18–35; Westwood, *Railways at War*, 8–12, 91, 197.

56. Showalter, *Railroads and Rifles*, 28.

57. Angevine, *Railroad and the State*, 142.

58. Bruce Waller, *Bismarck* (New York: Blackwell, 1997), especially chap. 3; William Carr, *A History of Germany, 1815–1990* (London: Edward Arnold, 1991), chap. 4.

59. Showalter, *Railroads and Rifles*, chap. 1; Bucholz, *Moltke and the German Wars, 1864–1871*, 40.

60. Bucholz, *Moltke, Schlieffen, and Prussian War Planning*, 36.

61. Bucholz, *Moltke and the German Wars, 1864–1871*, 61–65.

62. Daniel Hughes, ed., *Moltke on the Art of War: Selected Writings* (Novato, Calif.: Presidio, 1993), 107.

63. J. F. C. Fuller, *War and Western Civilization, 1832–1932* (London: Duckworth, 1932), 99. Others long ago denied that Moltke ever made the statement. See citations in Jay Luvaas, *The Military Legacy of the Civil War: The European Inheritance* (Chicago: University of Chicago Press, 1959), 126n21. Moreover, even though Luvaas seems to agree with the conventional view—"In the absence of evidence to the contrary, we can only assume that Moltke was far too involved in the momentous events taking place in Europe to cast more than a cursory glance at the war in America"—he notes an exception for the military lessons of the railroads. Ibid., 126, 122–24. Also see Jay Luvaas, "A Prussian Observer with Lee," *Military Affairs* 21 (Autumn 1957): 105–17.

64. As Bucholz writes, "By autumn 1865 Moltke knew a great deal about the American Civil War. How could he not? Moltke was a professional solider. This was the biggest war—by far—going on in the world at the time . . . [Moltke had ample time] to understand and incorporate some of its novel principles of industrial mass warfare. And so he did." Bucholz, *Moltke and the German Wars*, 104.

65. Pratt, *Rise of Rail-Power*, 104, 122.

66. Bucholz, *Moltke and the German Wars, 1864–1871*, 71–73; Thomas J. Adriance, *Last Gaiter Button*, 42.

67. Showalter, *Railroads and Rifles*, chaps. 1–3.

68. Hajo Holborn, "The Prusso-German School: Moltke and the Rise of the General Staff," in *Makers of Modern Strategy from Machiavelli to the Nuclear Age*, ed. Peter Paret (Princeton: Princeton University Press, 1986), 281–95; Gunther E. Rothenberg, "Moltke, Schlieffen, and the Doctrine of Strategic Envelopment," in Paret, *Makers of Modern Strategy*, 296–325; Hughes, *Moltke on the Art of War*.

69. Hughes, *Moltke on the Art of War*, 108, 113.

70. Bucholz, *Moltke and the German Wars*, 77–81; Showalter, *Railroads and Rifles*, 48–49; Bucholz, *Moltke, Schlieffen, and Prussian War Planning*, 43.

71. Dennis Showalter, "Mass Multiplied by Impulsion: The Influence of Railroads on Prussian Planning for the Seven Weeks' War," *Military Affairs* 38 (April 1974): 62–67; Showalter, *Railroads and Rifles*, 52–53; Bucholz, *Moltke and the German Wars*, 106–10; Bucholz, *Moltke, Schlieffen, and Prussian War Planning*, 43–45; Holborn, "Prusso-German School," 288; Rothenberg, "Moltke, Schlieffen, and the Doctrine of Strategic Envelopment," 296–300.

72. See Showalter, "Mass Multiplied by Impulsion," 62–63.

73. "Otto von Bismarck: Letter to Minister von Manteuffel, 1856," *Modern History Sourcebook: Documents of German Unification, 1848–1871,* http://www.fordham.edu/hal sall/mod/germanunification.html.

74. Craig, *Battle of Königgrätz,* 27–29; Showalter, *Railroads and Rifles,* 52–56; Bucholz, *Moltke and the German Wars,* 103–19.

75. Craig, *Battle of Königgrätz,* 29–30; Showalter, "Mass Multiplied by Impulsion," 63; Bucholz, *Moltke and the German Wars,* 110–17; Wawro, *Austro-Prussian War,* 20–21.

76. Wawro, *Austro-Prussian War,* 50–53.

77. Showalter, *Railroads and Rifles,* 62; Wawro, *Austro-Prussian War,* 56.

78. Carr, *History of Germany,* chap. 4, Waller, *Bismarck,* chap. 3; Showalter, *Railroads and Rifles,* 104; Van Creveld, *Supplying War,* 87–90.

79. Quester, *Offense and Defense,* 79–80.

80. Van Evera, *Causes of War,* 238.

81. Quester, *Offense and Defense,* 11. See also Van Evera, *Causes of War,* chap. 7. On the causes of the cult of the offensive, including those factors not related to technological misperceptions, see Jack Snyder, *The Ideology of the Offensive: Military Decision Making and the Disasters of 1914* (Ithaca: Cornell University Press, 1984).

82. A. J. P. Taylor, *War by Time-Table: How the First World War Began* (London: Macdonald, 1969), 15–45, 119–21; Barbara W. Tuchman, *The Guns of August* (New York: Macmillan, 1962). Also see A. J. P. Taylor, *The First World War: An Illustrated History* (1963; repr., New York: Perigree Books, 1980), 20–21; B. H. Liddell Hart, *History of the First World War* (London: Faber, 1972), 22–27; L. C. F. Turner, *Origins of the First World War* (New York: W. W. Norton, 1970).

83. This consensus continues to be bolstered as new documents emerge from the archives of the former East Germany. The first important challenge to the inadvertent-war thesis and effort to establish clear German guilt is Fritz Fischer, *War of Illusions: German Policies from 1911 to 1914* (1969; repr., New York: W. W. Norton, 1975). Critiques of the role of tight mobilization schedules in the outbreak of war are Marc Trachtenberg, "The Meaning of Mobilization in 1914," *International Security* 15 (Winter 1990–91): 120–50; Jack S. Levy, "Preferences, Constraints, and Choices in July 1914," *International Security* 15 (Winter 1990–91): 151–86; and David Stevenson, "Militarization and Diplomacy in Europe before 1914," *International Security* 22 (Summer 1997): 125–61. Scott Sagan challenges the causes and consequences of the "cult of the offensive" in "1914 Revisited: Allies, Offense, and Instability," *International Security* 11 (Fall 1986): 151–75. Among more recent scholarship in political science, Dale Copeland challenges the inadvertent war thesis in his book *The Origins of Major War* (Ithaca: Cornell University Press, 2000), chaps. 3–4. Excellent recent summaries of the historical debate on the origins of World War I are Annika Mombauer, *The Origins of the First World War: Controversies and Consensus* (London: Longman, 2002); and Richard F. Hamilton and Holger H. Herwig, eds., *The Origins of World War I* (New York: Cambridge University Press, 2003).

84. Strachan, *To Arms,* 166–67.

85. Holger H. Herwig, "Germany," in *Origins of World War I,* ed. Hamilton and Herwig, 150–87, at 154.

86. See Terence Zuber, "The Schlieffen Plan Reconsidered," *War in History* 6 (1999): 262–305; and Terence Zuber, *Inventing the Schlieffen Plan: German War Planning, 1871–1914* (New York: Oxford University Press, 2002). The quotation is from Zuber, *Inventing the Schlieffen Plan,* 304. Zuber argues that the postwar German officers' promulgation of the myth of the Schlieffen Plan has unfortunately received further validation

from influential historians such as Gerhard Ritter, Paul Kennedy, Gordon Craig, Jehuda Wallach, L. C. F. Turner, Martin Kitchen, Arden Bucholz, and Holger Herwig. See list of citations in Zuber, *Inventing the Schlieffen Plan,* 50. Ritter's work is most directly undermined by the Zuber thesis. See ibid., 42–51; Gerhard Ritter, *The Schlieffen Plan: Critique of a Myth* (New York: Praeger, 1958); and Gerhard Ritter, *The Sword and the Scepter: The Problem of Militarism in Germany,* 4 vols. (Coral Gables, Florida: University of Miami Press, 1969).

87. Hew Strachan largely supports Zuber: "All the older literature needs to be revised in the light of Zuber," he writes. Strachan, *To Arms,* 166. But Holger Herwig dismisses Zuber's claims as "utterly misleading" in "Germany and the 'Short War' Illusion: Toward a New Interpretation?" *Journal of Military History* 66 (July 2002): 681–93, at 683. See also Herwig, "Germany," 151–52. Terence Holmes gives a fuller critique in Terence M. Holmes, "The Reluctant March on Paris: A Reply to Terence Zuber's 'The Schlieffen Plan Reconsidered,' " *War in History* 8 (2001): 208–32. Zuber and Holmes continue the debate in Terence Zuber, "Terence Holmes Reinvents the Schlieffen Plan," *War in History* 8 (2001): 468–76; Terence M. Holmes, "The Real Thing: A Reply to Terence Zuber's 'Terence Holmes Reinvents the Schlieffen Plan,' " *War in History* 9 (2002): 111–20; Terence Zuber, "Terence Holmes Reinvents the Schlieffen Plan—Again," *War in History* 10 (2003): 92–101; and Terence M. Holmes, "Asking Schlieffen: A Further Reply to Terence Zuber," *War in History* 10 (2003): 464–79. See also Robert T. Foley, "The Origins of the Schlieffen Plan," *War in History* 10 (2003): 222–32.

88. Strachan, *To Arms,* 167.

89. See Zuber, *Inventing the Schlieffen Plan,* 1–5, 33–35, 160–64, 220–63, 298–304. Strachan, *To Arms,* 163–80; Strachan, *First World War,* 41–46.

90. Strachan, *First World War,* 56.

91. The most favorable interpretation of the evidence for offense-defense theory is that, as Jack Levy argues, the tight mobilization plans of all the major powers before World War I, which depended on the quick and efficient use of railroads, constrained last-minute diplomatic efforts to prevent war, but were not a primary cause of the war. Jack S. Levy, "Preferences, Constraints, and Choices in July 1914," 151–86. However, Marc Trachtenberg convincingly rejects the notion that statesmen in July 1914 were overwhelmed by the military imperatives of mobilization and thus lost control of the situation. Trachtenberg, "Meaning of Mobilization in 1914," 212–15.

92. The metaphor is borrowed from Stevenson, "War by Timetable?" 163.

93. Tuchman, *Guns of August,* 72.

94. This paragraph reflects the current consensus among historians of World War I. See Mombauer, *Origins of the First World War,* 1–20, 175–224; Hamilton and Herwig, *Origins of World War I,* especially chaps. 5, 14–15; and Strachan, *To Arms,* 1–102. On Germany's desire to concede the first move to their adversaries see Trachtenberg, *Meaning of Mobilization in 1914,* 137–47.

Chapter 3. The Small Arms and Artillery Revolution

1. Stephen Van Evera, *Causes of War: Power and the Roots of Conflict* (Ithaca: Cornell University Press, 1999), 193–94. See also Stephen Van Evera, "The Cult of the Offensive and the Origins of the First World War," *International Security* 9 (Summer 1984): 58–107.

2. Robert Jervis, "Cooperation Under the Security Dilemma," *World Politics* 30 (January 1978): 167–214, at 191.

3. George H. Quester, *Offense and Defense in the International System* (New York: John Wiley & Sons, 1977), 101. On the "short war illusion" see Lancelot L. Farrar, *The Short-War Illusion: German Policy, Strategy and Domestic Affairs, August–December 1914* (Santa Barbara, Calif.: ABC-Clio, 1973).

4. James Fearon, "The Offense-Defense Balance and War Since 1648" (paper presented at the International Studies Association annual meeting, Chicago, February 21–25, 1995), 2; Richard K. Betts, "Must War Find a Way? A Review Essay," *International Security* 24 (Fall 1999): 166–98, at 184.

5. Some proponents of the "cult of the offensive" thesis do not rely on the argument that leaders misperceived the nature of prevailing technology. For example, Jack Snyder acknowledges that Europeans foresaw the effects of defensive firepower but pursued offensive strategies because such strategies served their organizational interests and parochial outlook. Jack Snyder, "Civil-Military Relations and the Cult of the Offensive, 1914 and 1984," *International Security* 9 (Summer 1984): 108–46. Others emphasize the extent to which European militaries tried to inculcate higher morale on the offensive as a means of overcoming the superiority of defensive firepower. Timothy Travers notes the British army's growing awareness of defensive firepower: "Yet the army generally sought to overcome these technological factors by reemphasizing the value of moral qualities and the use of those qualities in the offensive, thus attempting to answer technical problems with moral solutions. "Technology, Tactics, and Morale: Jean De Bloch, the Boer War, and British Military Theory, 1900–1914," *Journal of Modern History* 51 (June 1979), 264–86, at 264. In France, the army emphasized the offensive spirit not just as a means to cope with the challenges created by modern firepower but also as a solution for dealing with domestic civil-military and manpower problems. Douglas Porch writes, "The only way to cope with the new technical developments despite poor French resources was to rely on the patriotic audacity of French soldiers." *The March to the Marne: The French Army, 1871–1914* (Cambridge: Cambridge University Press, 1981), chap. 11. Despite his claim that Europeans overlooked the power of new defensive technologies, Stephen Van Evera's own discussion occasionally reveals that military strategists did not ignore, but sought ways to *overcome*, defensive firepower. See *Causes of War,* 197.

6. For overviews of these developments see Bernard Brodie and Fawn M. Brodie, *From Crossbow to H-Bomb* (Bloomington: Indiana University Press, 1973), 124–31; and T. N. Dupuy, *The Evolution of Weapons and Warfare* (Indianapolis: Bobbs-Merrill, 1980), chaps. 19–23; William McNeil, *The Pursuit of Power: Technology, Armed Force, and Society since A.D. 1000* (Chicago: University of Chicago Press, 1982), 223–306; and Hew Strachan, *European Armies and the Conduct of War* (London: Routledge, 1983), 108–49. In general, the most significant advances in small arms occurred in the first half of this period, while the greatest artillery developments emerged in the second half. Machine guns wrought their effects mainly after 1900.

7. Trevor Royle, *Crimea: The Great Crimean War, 1854–1856* (New York: St. Martin's Press, 2000), pt. 2; Winfried Baumgart, *The Crimean War, 1853–1856* (New York: Oxford University Press, 1999), chaps. 11–16; Larry H. Addington, *The Patterns of War since the Eighteenth Century* (Bloomington: Indiana University Press, 1994), 62–65; Strachan, *European Armies,* 114.

8. David S. Heidler and Jeanne T. Heidler, eds., *Encyclopedia of the American Civil War: A Political, Social, and Military History* (New York: W. W. Norton, 2000), 1649–50, 1802–3.

9. This battle in particular made the Union forces wary of such kinds of frontal assaults, inflicting them with what amounted to "Cold Harbor Syndrome." Ibid., 466.

10. Arden Bucholz, *Moltke and the German Wars, 1864–1871* (New York: Palgrave, 2001), 81–97.

11. Dennis E. Showalter, *Railroads and Rifles: Soldiers, Technology, and the Unification of Germany* (Hamden, Conn.: Archon, 1975), chap. 7; Geoffrey Wawro, *The Austro-Prussian War: Austria's War with Prussia and Italy in 1866* (New York: Cambridge University Press, 1996); Richard Holmes, ed., *The Oxford Companion to Military History* (New York: Cambridge University Press, 2001), 114.

12. Geoffrey Wawro, *The Franco-Prussian War: The German Conquest of France in 1870–1871* (New York: Cambridge University Press, 2003), 169–85; Michael Howard, *The Franco-Prussian War: the German Invasion of France, 1870–1871* (New York: Macmillan, 1962), chap. 5.

13. Dennis Showalter, "Prussia, Technology, and War: Artillery From 1815 to 1914," in *Men, Machines, and War,* ed. Ronald Haycock and Keith Neilson (Waterloo, Ont.: Wilfried Laurier University Press, 1988), 113–51; Michael Howard, *War in European History* (Oxford: Oxford University Press, 1976), 102–3.

14. Brodie and Brodie, *From Crossbow to H-Bomb,* 138–39; R. Ernest Dupuy and Trevor N. Dupuy, *The Harper Encyclopedia of Military History: From 3500 B.C. to the Present* (New York: HarperCollins, 1993), 925.

15. Thomas Pakenham, *The Boer War* (New York: Random House, 1994); Michael Howard, "Men against Fire: Expectations of War in 1914," *International Security* 9 (Summer 1984): 41–58, at 45–47; Dupuy and Dupuy, *Encyclopedia,* 933–36.

16. Addington, *Patterns of War,* 131; Howard, "Men against Fire," 52–54.

17. David Stevenson, *Cataclysm: The First World War as Political Tragedy* (New York: Basic Books, 2004), chap. 6; Hew Strachan, *The First World War* (New York: Viking, 2003), 184–97; Martin Middlebrook, *The First Day on the Somme, 1 July 1916* (New York: W. W. Norton, 1972); Timothy Travers, *The Killing Ground: The British Army, the Western Front and the Emergence of Modern Warfare, 1900–1918* (London: Allen & Unwin, 1987), chaps. 6–7.

18. See Gunther E. Rothenberg, "Moltke, Schlieffen, and the Doctrine of Strategic Envelopment," in *Makers of Modern Strategy from Machiavelli to the Nuclear Age,* ed. Peter Paret (Princeton: Princeton University Press, 1986), 296–325; Wawro, *Franco-Prussian War,* chap. 9.

19. Dupuy, *Evolution of Weapons and Warfare,* 201–2; Dupuy and Dupuy, *Encyclopedia,* 961–62, 979–82, and 990–91.

20. Howard, "Men against Fire," 46–47.

21. Dupuy and Dupuy, *Encyclopedia,* 933–36.

22. Martin van Creveld, *Technology and War: From 2000 B.C. to the Present* (New York: Free Press, 1991), 177.

23. Scott Sagan makes this point in arguing that proponents of the "cult of the offensive" have exaggerated the degree to which offensive operations were doomed to fail. See Scott D. Sagan, "1914 Revisited: Allies, Offense, and Instability," *International Security* 11 (Fall 1986): 151–75, at 158–61.

24. On the latest debate on the Schlieffen Plan, see Terence Zuber, "The Schlieffen Plan Reconsidered," *War in History* 6 (1999): 262–305; Terence M. Holmes, "The Reluctant March on Paris: A Reply to Terence Zuber's 'The Schlieffen Plan Reconsidered,' " *War in History* 8 (2001): 208–32. The debate among Zuber, Holmes, and others continues in subsequent issues of the journal. See also Terence Zuber, *Inventing the Schlieffen Plan: German War Planning, 1871–1914* (New York: Oxford University Press, 2002).

25. Martin van Creveld, *Supplying War: Logistics from Wallenstein to Patton* (London: Cambridge University Press, 1977), chap. 4; Gerhard Ritter, *The Schlieffen Plan: Critique of a Myth* (New York: Praeger, 1958); John Keegan, *The First World War* (New York: Alfred A. Knopf, 1999), chap. 2.

26. On Moltke's alleged adulteration of the Schlieffen Plan and a convincing rebuttal of that argument see Annika Mombauer, *Helmuth Von Moltke and the Origins of the First World War* (Cambridge: Cambridge University Press, 2001). Also see Hew Strachan, *The First World War: To Arms* (Oxford: Oxford University Press, 2001), 163–80; and Strachan, *First World War*, 40–48.

27. See Norman Stone, *The Eastern Front, 1914–1917* (New York: Charles Scribner's Sons, 1975); Dennis E. Showalter, *Tannenberg: Clash of Empires* (Hamden, Conn.: Archon Books, 1991); and Spencer C. Tucker, *The Great War, 1914–18* (Bloomington: Indiana University Press, 1998.

28. Howard, "Men against Fire," 54.

29. Timothy T. Lupfer, *The Dynamics of Doctrine: The Changes in German Tactical Doctrine During the First World War* (Fort Leavenworth, Kans.: U.S. Army Command and General Staff College, 1981); Dupuy, *Evolution of Weapons and Warfare*, 225–29; and Strachan, *European Armies*, 142–49.

30. The Russian Brusilov offensive in June 1916 was an earlier example of the tactical breakthroughs that were possible when infantry was used intelligently and in cooperation with artillery support. See Strachan, *European Armies*, 143–44; and Christopher Bellamy, *The Evolution of Modern Land Warfare: Theory and Practice* (London: Routledge, 1990), chap. 3.

31. Stephen Biddle, *Military Power: Explaining Victory and Defeat in Modern Battle* (Princeton: Princeton University Press, 2004), chap. 5. Biddle argues that the "stable and essentially transnational body of ideas on the methods needed to operate effectively in the face of radically lethal modern weapons . . . broke the trench stalemate in 1918 and defined the standard for successful military operations throughout the post-1918 era." He labels this body of knowledge the "modern system of force employment." Ibid., 28.

32. See Paddy Griffith, *Battle Tactics of the Western Front: The British Army's Art of Attack, 1916–18* (New Haven: Yale University Press, 1994); John Terraine, *To Win a War: 1918, The Year of Victory* (London: Sidgwick and Jackson, 1978). Timothy Travers argues, somewhat differently, that for the British it was more and better armaments rather than improved tactics that made the difference in 1918. See Timothy Travers, *How the War Was Won: Command and Technology in the British Army on the Western Front, 1917–1918* (New York: Routledge, 1992).

33. Van Evera, *Causes of War*, 193–94. Van Evera also claims, "Belief in the strength of the offense peaked in 1914 in Europe, and Germany had the largest offensive opportunities and defensive vulnerabilities among Europe's powers. Offense-defense theory therefore forecasts that war should erupt in Europe in about 1914, authored largely by

Germany." Ibid., 199n25. However, this is essentially a nonfalsifiable claim. If war had instead broken out in 1890, 1905, or 1912, for example, Van Evera would still be able to claim theory validation. That is, if beliefs in offense were steadily rising from 1890, these beliefs were potentially always at their "peak" whenever war broke out.

34. Quester, *Offense and Defense*, 11.

35. John H. Maurer, *The Outbreak of the First World War: Strategic Planning, Crisis Decision Making, and Deterrence Failure* (Westport, Conn.: Praeger, 1995), xi. See also Farrar, *Short-War Illusion*.

36. See Stig Förster, "Dreams and Nightmares: German Military Leadership and the Images of Future Warfare, 1871–1914," in *Anticipating Total War: The German and American Experiences, 1871–1914*, ed. Manfred F. Boemeke, Roger Chickering, and Stig Förster (Washington, D.C.: German Historical Institute, 1999), 343–76. For a short summary of the Förster thesis see Holger H. Herwig, "Germany and the 'Short War' Illusion: Toward a New Interpretation?" *Journal of Military History* 66 (July 2002), 681–94. Also see Strachan, *To Arms*, 1005–14; Mombauer, *Helmuth Von Moltke*, 211, 283–89.

37. On the consensus view among historians about the origins of World War I and the degree of German culpability see Annika Mombauer, *The Origins of the First World War: Controversies and Consensus* (London: Longman, 2002); and Richard F. Hamilton and Holger H. Herwig, eds., *The Origins of World War I* (New York: Cambridge University Press, 2003), especially chap. 5.

38. See Arden Bucholz, *Moltke, Schlieffen, and Prussian War Planning* (New York: Berg, 1991); Bucholz, *Moltke and the German Wars;* Larry H. Addington, *The Blitzkrieg Era and the German General Staff, 1865–1941* (New Brunswick, N.J.: Rutgers University Press, 1971); Hajo Holborn, "The Prusso-German School: Moltke and the Rise of the General Staff," in Paret, *Makers of Modern Strategy*, 281–95, at 290–91; and Rothenberg, "Moltke, Schlieffen, and the Doctrine of Strategic Envelopment," 296–325, at 301–2.

39. Showalter, *Railroads and Rifles*, 99.

40. Bucholz, *Moltke and the German Wars*, 72.

41. Ibid., 63, 106–8.

42. Helmuth von Moltke, "Defensive Position, Envelopment, and Base of Operations (1865)," in *Moltke on the Art of War: Selected Writings*, ed. Daniel Hughes (Novato, Calif.: Presidio, 1993), 49, 51.

43. Helmuth von Moltke, "Instructions for Large Unit Commanders of 24 June 1869," in Hughes, *Moltke on the Art of War*, 201–2, 216.

44. Showalter, *Railroads and Rifles*, 101 (emphasis in original).

45. Eric Dorn Brose, *The Kaiser's Army: The Politics of Military Technology in Germany During the Machine Age, 1870–1918* (New York: Oxford University Press, 2001).

46. Ritter, *Schlieffen Plan*, 18.

47. Helmuth von Moltke, "Defensive and Offensive (1874)," in Hughes, *Moltke on the Art of War*, 52.

48. Rothenberg, "Moltke, Schlieffen, and the Doctrine of Strategic Envelopment," 306–11; Zuber, *Inventing the Schlieffen Plan*, 52–134.

49. Michael Geyer, "German Strategy in the Age of Machine Warfare, 1914–1945," in Paret, *Makers of Modern Strategy*, 527–97, at 530; Walter Goerlitz, *History of the German General Staff, 1657–1945* (Boulder, Colo.: Westview Press, 1985), 102.

50. The most radical claims, including the argument that "the Schlieffen Plan never existed" are found in Zuber, *Inventing the Schlieffen Plan*. One critical response is Terence M. Holmes, "The Reluctant March on Paris: A Reply to Terence Zuber's 'The

Schlieffen Plan Reconsidered,' " *War in History* 8 (2001): 208–32. Some additional discussion of this debate and a list of citations are found in chapter 2.

51. Förster, "Dreams and Nightmares," 355.

52. Ritter, *Schlieffen Plan,* 144.

53. Ibid., 145.

54. Strachan, *To Arms,* 163–72, 1008–10; Zuber, *Inventing the Schlieffen Plan,* chap. 4; Förster, "Dreams and Nightmares"; Mombauer, *Helmuth von Moltke,* 72–80, 95; Ritter, *Schlieffen Plan,* 145–95.

55. See Förster, "Dreams and Nightmares"; Herwig, "Germany and the 'Short War' Illusion," 688–92; Mombauer, *Helmuth Von Moltke,* 94–95, 211, 285, 287–88.

56. Strachan, *To Arms,* 172.

57. Förster, "Dreams and Nightmares," 363–64.

58. Herwig, "Germany and the "Short War" Illusion," 688–89.

59. Geyer, "German Strategy in the Age of Machine Warfare," 530–31.

60. Mombauer, *Helmuth Von Moltke,* 95.

61. Förster, "Dreams and Nightmares," 364–65.

62. Herwig, "Germany and the "Short War" Illusion," 691.

63. Helmuth von Moltke, "General Observations on the Schlieffen Plan," in Ritter, *Schlieffen Plan,* 165.

64. Mombauer, *Helmuth Von Moltke,* 93–98.

65. Förster, "Dreams and Nightmares," 364–65, 373; Herwig, "Germany and the "Short War" Illusion," 692.

66. See Mombauer, *Helmuth Von Moltke,* 283–89. She writes, "The evidence now available confirms without a doubt that Moltke and his colleagues wanted war and had sufficient influence over their political colleagues to achieve their aim." Ibid., 287.

67. Strachan identifies two additional factors: the existence of large and powerful alliances and the impact of universal military service and national mobilization. *To Arms,* 1007–8.

68. Howard, "Men against Fire," 43. See also Howard, *War in European History,* 104–7; Travers, "Technology, Tactics, and Morale"; Travers, *Killing Ground,* chap. 3; Theodore Ropp, *War in the Modern World* (New York: Collier Books, 1962), 222–27; and Strachan, *European Armies,* chaps. 8–9.

Chapter 4. The Armored Revolution

1. Stephen Van Evera, *Causes of War: Power and the Roots of Conflict* (Ithaca: Cornell University Press, 1999), 162.

2. Robert Jervis, "Cooperation under the Security Dilemma," *World Politics* 30 (January 1978): 167–214, at 197. See Quincy Wright, *A Study of War* (Chicago: University of Chicago Press, 1965), 142.

3. Sean M. Lynn-Jones, "Offense-Defense Theory and Its Critics," *Security Studies* 4 (Summer 1995): 660–91, at 676.

4. Charles L. Glaser and Chaim Kaufmann, "What Is the Offense-Defense Balance and Can We Measure It?" *International Security* 22 (Spring 1998): 44–82, at 63.

5. Van Evera, *Causes of War,* 123, 175, 177. Consistent with his broad conception of the offense-defense balance, he notes that Hitler's faith in the offensive was driven by more than just technology, including his racist and social Darwinist views. Neverthe-

less, Van Evera's discussion makes it clear that the most important dimension of Hitler's belief in offense dominance was his view of the tank as an offensive innovation.

6. Ibid., 175.

7. On the role of tanks in World War I, see R. M. Ogorkiewicz, *Armoured Forces: A History of Armoured Forces and Their Vehicles* (New York: Arco, 1970); Williamson Murray, "Armored Warfare: The British, French, and German Experiences," in *Military Innovation in the Interwar Period*, ed. Williamson Murray and Allan R. Millett (Cambridge: Cambridge University Press, 1996), 6–49; J. P. Harris, *Men, Ideas, and Tanks: British Military Thought and Armoured Forces, 1903–1939* (Manchester: Manchester University Press, 1995); and Spencer C. Tucker, *The Great War, 1914–18* (Bloomington: Indiana University Press, 1998), 109–11, 141–43, 170–71. On infantry tactics, see Timothy T. Lupfer, *The Dynamics of Doctrine: The Changes in German Tactical Doctrine During the First World War* (Fort Leavenworth, Kans.: U.S. Army Command and General Staff College, 1981); T. N. Dupuy, *The Evolution of Weapons and Warfare* (Indianapolis: Bobbs-Merrill, 1980), 225–29; Tucker, *The Great War,* 159–62; and Hew Strachan, *European Armies and the Conduct of War* (London: Routledge, 1983), 142–49.

8. For an overview, see Archer Jones, *The Art of War in the Western World* (New York: Oxford University Press, 1987), 497–507; and Larry H. Addington, *The Patterns of War Since the Eighteenth Century* (Bloomington: Indiana University Press, 1994), 191–94.

9. Mary R. Habeck, *Storm of Steel: The Development of Armor Doctrine in Germany and the Soviet Union, 1919–1939* (Ithaca: Cornell University Press, 2003), chap. 7.

10. Edward J. Drea, *Nomonhan: Japanese-Soviet Tactical Combat, 1939* (Fort Leavenworth, Kans.: U.S. Army Command and General Staff College, 1981); Alvin D. Coox, *Nomonhan: Japan against Russia, 1939*, 2 vols. (Stanford, Calif.: Stanford University Press, 1985), especially chaps. 21–26, 30–36.

11. Williamson Murray and Allan R. Millett, *A War to Be Won: Fighting in the Second World War* (Cambridge, Mass.: Belknap Press, 2000), 47–48.

12. J. P. Harris, "The Myth of Blitzkrieg," *War in History* 2 (1995): 343.

13. See Matthew Cooper, *The German Army, 1933–1945: Its Political and Military Failure* (New York: Stein and Day, 1978), 169–76; B. H. Liddell Hart, *History of the Second World War* (New York: G. P. Putnam's Sons, 1971), 27–32; and Jones, *Art of War,* 508–9.

14. Germany deployed 2,439 tanks against 3,079 French and British tanks. If one includes Belgian and Dutch tanks, the Allies deployed a total of 4,204 tanks in May 1940. In manpower terms, Germany deployed 102 divisions against 123 Allied divisions. Ernest R. May, *Strange Victory: Hitler's Conquest of France* (New York: Hill and Wang, 2000), 477–78.

15. Julian Jackson, *The Fall of France: The Nazi Invasion of 1940* (New York: Oxford University Press, 2003), 12–17; May, *Strange Victory,* 477–78.

16. For the ledger of German strengths and allied weaknesses, see May, *Strange Victory,* 448–64; Jones, *Art of War,* 510–44; Larry H. Addington, *The Blitzkrieg Era and the German General Staff, 1865–1941* (New Brunswick, N.J.: Rutgers University Press, 1971), 101–23; Cooper, *German Army,* 214–15; Liddell Hart, *History of the Second World War,* chap. 7; Robert Allan Doughty, *The Breaking Point: Sedan and the Fall of France, 1940* (Hamden, Conn.: Archon, 1990), chap. 1; and Jackson, *Fall of France.*

17. Liddell Hart, *History of the Second World War,* 707–8.

18. Moreover, it is worth noting that in the first major tank battle of the war, in northern France near the Dyle River, the German armored penetration was halted. In reference to this battle, Archer Jones notes that "defense exhibited its predominance in combat between the same weapon systems." *Art of War,* 525.

19. On the Red Army's deficiencies on the eve of the war, see Alan Clark, *Barbarossa: The Russian-German Conflict, 1941–45* (New York: William Morrow, 1965), chap. 2.

20. Ibid., 34. See also John Erickson, *The Soviet High Command: A Military-Political History, 1918–1941* (London: St. Martin's Press, 1962).

21. Timothy A. Wray, *Standing Fast: German Defensive Doctrine on the Russian Front During World War II, Prewar to March 1943* (Fort Leavenworth, Kans.: U.S. Army Command and General Staff College, 1986), 25.

22. Ibid., 35.

23. Habeck, *Storm of Steel*, xvii.

24. Glaser and Kaufmann, "What Is the Offense-Defense Balance?" 56.

25. See Habeck, *Storm of Steel*.

26. For an account of the Kharkov offensive, see David M. Glantz and Jonathan M. House, *When Titans Clashed: How the Red Army Stopped Hitler* (Lawrence: University Press of Kansas, 1995), 111–16; and Clark, *Barbarossa*, 198–203.

27. Glantz and House, *When Titans Clashed*, 301–5 (Table C).

28. Wray, *Standing Fast*, 121; Also, see Glantz and House, *When Titans Clashed*, 152–53; and Cooper, *German Army*, 460–79.

29. Heinz Guderian, *Panzer Leader* (New York: De Capo Press, 1996), 287–300; Glantz and House, *When Titans Clashed*, 153.

30. For example, Field Marshal Erich von Manstein recounts the following conversation with Hitler: " 'One thing we must be clear about, mein Fuhrer,' I began, 'is that the extremely critical situation we are now in cannot be put down to the enemy's superiority alone, great though it is. It is also due to the way in which we are led.' Erich Von Manstein, *Lost Victories*, ed. and trans. Anthony G. Powell (Novato, Calif.: Presidio Press, 1982), 504. Also see F. W. Von Mellenthin, *Panzer Battles: A Study of the Employment of Armor in the Second World War* (New York: Ballantine Books, 1956); and Guderian, *Panzer Leader*.

31. See H. P. Willmott and John Keegan, *The Second World War in the East* (London: Cassell, 1999); and Murray and Millett, *A War to Be Won*. See also the discussion of how the Red Army evolved to the point where it could conduct deep offensive operations in Glantz and House, *When Titans Clashed*, 154–57. Also see George F. Hofmann, "Doctrine, Tank Technology, and Execution: I. A. Khalepskii and the Red Army's Fulfillment of Deep Offensive Operations," *Journal of Slavic Military Studies* 9 (June 1996): 283–334.

32. For a thorough discussion of German defensive doctrine in the east, see Wray, *Standing Fast*. The ideal defense in depth would have an echeloned network of prepared defensive positions designed to wear an attacker down as the attacker advances. John J. Mearsheimer, *Conventional Deterrence* (Ithaca: Cornell University Press, 1983), 49.

33. Quoted in Wray, *Standing Fast*, 170.

34. Guderian, *Panzer Leader*, 328.

35. Wray, *Standing Fast*, 118.

36. Glantz and House, *When Titans Clashed*, 136–39.

37. Wray, *Standing Fast*, 148–51.

38. For a detailed account of these operations in the Donbas and Kharkov regions, see David M. Glantz, *From the Don to the Dnepr: Soviet Offensive Operations, December 1942–August 1943* (London: Frank Cass, 1991); Von Manstein, *Lost Victories;* Von Mellenthin, *Panzer Battles;* and Wray, *Standing Fast*, 155–64.

39. Wray, *Standing Fast*, 155.

40. Albert Seaton, *The Fall of Fortress Europe* (New York: Holmes & Meier, 1981), 55.

41. Glantz and House, *When Titans Clashed,* 167.

42. Ibid., 184.

43. Ibid., 174–75.

44. For broader discussion of the origins of World War II, Hitler's goals, and Nazi Germany's foreign policy see P. M. H. Bell, *The Origins of the Second World War in Europe* (London: Longman, 1986); Andreas Hillgruber, *Germany and the Two World Wars,* trans. William C. Kirby (Cambridge: Harvard University Press, 1981); Gerhard L. Weinberg, *The Foreign Policy of Hitler's Germany: Diplomatic Revolution in Europe, 1933–36* (Chicago: University of Chicago Press, 1970); and Gerhard L. Weinberg, *The Foreign Policy of Hitler's Germany: Starting World War II, 1937–1939* (Chicago: University of Chicago Press, 1980).

45. For example, see Weinberg, *Foreign Policy of Hitler's Germany,* 662–63.

46. Randall L. Schweller, *Deadly Imbalances: Tripolarity and Hitler's Strategy of World Conquest* (New York: Columbia University Press, 1998), chap. 4.

47. Weinberg, *Foreign Policy of Hitler's Germany,* 463.

48. Donald Cameron Watt, *How War Came: The Immediate Origins of the Second World War, 1938–1939* (London: Pimlico, 2001), 623–24.

49. See Harris, *Men, Ideas, and Tanks;* and Brian Bond, *British Military Policy Between the Two World Wars* (New York: Oxford University Press, 1980).

50. The quotation is from Fuller's winning submission for the 1919 Royal United Service Institution's Gold Medal essay competition, cited in Harris, *Men, Ideas, and Tanks,* 203.

51. Murray, "Armored Warfare," 19–29; Harris, *Men, Ideas, and Tanks,* chaps. 6–8.

52. On Fuller, see J. F. C. Fuller, *The Reformation of War* (New York: E. P. Dutton, 1923), 152–69; Brian Holden Reid, *J. F. C. Fuller: Military Thinker* (New York: St. Martin's, 1987), 57, 183, and chap. 7; and Harris, *Men, Ideas, and Tanks,* chaps. 6–8. On Liddell Hart, see John J. Mearsheimer, *Liddell Hart and the Weight of History* (Ithaca: Cornell University Press, 1988), 36–45, 105–23.

53. In May 1940, the heaviest French tank—the Char B1—had both a 75mm and a 47mm cannon, along with 60mm thick armor. The heaviest German tanks—Panzer IIIs and Panzer IVs—had single 37mm and single 75mm (short barrel) cannons, respectively, and 30mm armor. Although Germany had more light tanks (1,478 vs. 765), France possessed greater numbers of medium tanks (1,400 vs. 961) and heavy tanks (584 vs. 0). As many as half of Germany's basic tanks (Panzer Is and IIs) broke down in Poland. May, *Strange Victory,* 209, 478.

54. Robert Allan Doughty, *The Seeds of Disaster: The Development of French Army Doctrine, 1919–1939* (Hamden, Conn.: Archon Books, 1985).

55. See Cooper, *German Army;* Barry R. Posen, *The Sources of Military Doctrine: France, Britain, and Germany between the World Wars* (Ithaca: Cornell University Press, 1984); Michael Geyer, "German Strategy in the Age of Machine Warfare, 1914–1945," in *Makers of Modern Strategy from Machiavelli to the Nuclear Age,* ed. Peter Paret (Princeton: Princeton University Press, 1986), 527–97; and James S. Corum, *The Roots of Blitzkrieg, Hans Von Seeckt and German Military Reform* (Lawrence: University of Kansas, 1992).

56. Cooper, *German Army,* 133.

57. Murray, "Armored Warfare," 34–38.

58. For this and other influential factors, see Posen, *Sources of Military Doctrine,* 182–88.

59. Cooper, *German Army*, 132–37.

60. Addington, *Blitzkrieg Era*, 29–30; and Peter Calvocoressi, Guy Wint, and John Pritchard, *Total War: The Causes and Courses of the Second World War*, rev. 2d ed. (New York: Pantheon Books, 1989), 130–31.

61. Wray, *Standing Fast*, 6.

62. Harris, *Men, Ideas, and Tanks*, 219, 228.

63. Wray, *Standing Fast*, 17–18; Harris, *Men, Ideas, and Tanks*, 228; and Mearsheimer, *Liddell Hart*, 80.

64. Harris, "The Myth of Blitzkrieg," 335–52.

65. Cooper, *German Army*, 148–58. See also Guderian, *Panzer Leader;* and Posen, *Sources of Military Doctrine*, 205–19.

66. Habeck, *Storm of Steel*, 273, 281–83.

67. See Heinz Guderian, *Achtung—Panzer!*, trans. Christopher Duffy (London: Arms and Armour Press, 1992), 188–90. Noted in Habeck, *Storm of Steel*, 252.

68. Harris, "The Myth of Blitzkrieg"; Cooper, *German Army;* J. P. Harris and F. H. Toase, *Armoured Warfare* (New York: St. Martin's, 1990), 64–69.

69. Van Evera, *Causes of War*, 123. The argument is presented at greater length in Mearsheimer, *Conventional Deterrence*, 99–133.

70. See Cooper, *German Army*, 179–215; Doughty, *Breaking Point*, 19–25.

71. May, *Strange Victory*, 18.

72. Mearsheimer, *Conventional Deterrence*, 111.

73. May, *Strange Victory*, 15–27, 218–20, 254.

74. Ibid., 225.

75. John Mearsheimer argues that Hitler's original plan was thwarted "mainly by the determined opposition of his generals . . . without their support, Hitler was unable to strike in the West." Mearsheimer, *Conventional Deterrence*, 131–32. Given Hitler's strong determination to fight the Allies at the earliest opportunity, his willingness to take great gambles, his resistance to genuine collaboration in the strategic decision-making process, and his inflated confidence in his own leadership, this argument is unpersuasive. It seems more reasonable to conclude that Hitler would have proceeded even without this strange turn of events.

76. Some argue that even the attack launched in May 1940 was based on essentially traditional operational principles and methods, and was not motivated by the belief that armored forces had transformed warfare. See Harris, "The Myth of Blitzkrieg"; Cooper, *German Army;* Harris and Toase, *Armoured Warfare*, 64–69; and Doughty, *Breaking Point*, 323. Doughty argues, for example, that "the evolution of the German plan suggests that the purpose of the phalanx of forces moving through the Ardennes was the traditional *kesselschlacht* strategy of encirclement and annihilation." Doughty, *Breaking Point*, 323. The key debate centers on whether the German strategy primarily sought to attack the Allied "central nervous system," which is the fundamentally distinguishing characteristic of a blitzkrieg strategy, or rather simply aimed at a traditional envelopment and decisive battle. Harris writes,

> [Heinz] Guderian did not recommend the use of infiltration tactics by tanks, nor the singling out of major headquarters for special attention by armoured forces once the initial breakthrough had been achieved. He did not state that the main purpose of armoured breakthrough was to bring about a collapse of the enemy's command and control, nor that such a collapse was the principal means by which

the enemy's defeat was to be achieved. Doubtless Guderian did expect some break-down of command and control to occur as a result of a major armoured break-through; but it seems more likely that he was relying mainly on the physical rather than the psychological effects of breakthrough, on manoeuvering the enemy into an impossible position rather than causing his command system to fail.

"The Myth of Blitzkrieg," 342. Ultimately, it is difficult to know for sure how the Ger-man penetration was to proceed because the final operation order for the attack did not stipulate what the armored forces would do in the event of a successful break-through.

77. May, *Strange Victory,* 7, 267.

Chapter 5. The Nuclear Revolution

1. Bernard Brodie, ed., *The Absolute Weapon: Atomic Power and World Order* (New York: Harcourt, Brace, 1946), 52.

2. A prominent exception is Kenneth N. Waltz, *The Spread of Nuclear Weapons: More May Be Better,* Adelphi Papers no. 171 (London: International Institute for Strategic Studies, 1981); and Waltz, "Nuclear Myths and Political Realities," *American Political Sci-ence Review* 84 (September 1990): 731–45; and Waltz's contributions in Scott D. Sagan and Kenneth N. Waltz, *The Spread of Nuclear Weapons: A Debate Renewed* (New York: W. W. Norton, 2003).

3. As Richard Betts writes, "If the seedbed of deterrence theory does not account for the production of recent ODT, it certainly accounts for the intellectual receptivity of its consumers in the peak period of attention in the 1980s." Richard K. Betts, "Must War Find a Way? A Review Essay," *International Security* 24 (Fall 1999): 166–98, at 176. The close connection between nuclear deterrence theory and offense-defense theory is found in George H. Quester, *Offense and Defense in the International System* (New York: John Wiley, 1977), chap. 13; Robert Jervis, "Cooperation under the Security Dilemma," *World Politics* 30 (January 1978): 167–214; Shai Feldman, *Israeli Nuclear Deterrence* (New York: Columbia University Press, 1982); Robert Jervis, *The Illogic of American Nuclear Strategy* (Ithaca: Cornell University Press, 1984); Jack Snyder, *The Ideology of the Offensive: Military Decision Making and the Disasters of 1914* (Ithaca: Cornell University Press, 1984); Robert Jervis, *The Meaning of the Nuclear Revolution: Statecraft and the Prospect of Nuclear Armageddon* (Ithaca: Cornell University Press, 1989); Charles L. Glaser, *Analyz-ing Strategic Nuclear Policy* (Princeton, N.J.: Princeton University Press, 1990); Stephen Van Evera, *Causes of War: Power and the Roots of Conflict* (Ithaca: Cornell University Press, 1999), chap. 8. Although Kenneth Waltz is not typically seen as an offense-defense the-orist, see also Kenneth N. Waltz, *Theory of International Politics* (Reading, Mass.: Addison-Wesley, 1979); "Spread of Nuclear Weapons"; "Nuclear Myths and Political Re-alities"; and Sagan and Waltz, *Spread of Nuclear Weapons.*

4. Thus the Robert Jervis book is titled *The* Illogic *of American Nuclear Strategy.*

5. Van Evera, *Causes of War,* 162.

6. Charles L. Glaser, "When Are Arms Races Dangerous? Rational versus Suboptimal Arming," *International Security* 28 (Spring 2004): 44–84, at 75. Other nuclear analysts disagree. See Beckman and others, *The Nuclear Predicament: Nuclear Weapons in the Twenty-First Century,* 3rd ed. (Upper Saddle River, N.J.: Prentice Hall, 2000), 311

7. Thomas C. Schelling, *Arms and Influence* (New Haven: Yale University Press, 1966), 19.

8. Jervis, "Cooperation under the Security Dilemma," 198.

9. Glaser, *Analyzing Strategic Nuclear Policy,* 94–99; Jervis, "Cooperation under the Security Dilemma," 198; Van Evera, *Causes of War,* 177–78, 246–47. On the related concepts of deterrence by denial and deterrence by punishment, see Glenn H. Snyder, *Deterrence and Defense: Toward a Theory of National Security* (Princeton, N.J.: Princeton University Press, 1961).

10. For example, see George N. Lewis, Theodore A. Postol, and John Pike, "Why National Missile Defense Won't Work," *Scientific American,* August 1999, 36–41.

11. Van Evera, *Causes of War,* 177.

12. See Betts, "Must War Find a Way?" 179–80.

13. Jervis, *Meaning of the Nuclear Revolution,* 8.

14. Glenn Snyder, "The Balance of Power and the Balance of Terror," in *The Balance of Power,* ed. Paul Seabury (San Francisco: Chandler, 1965), 184–201. Also see Jervis, *Illogic of American Nuclear Strategy,* 29–34.

15. Jervis, *Meaning of the Nuclear Revolution,* 19–23.

16. See John Lewis Gaddis, *The Long Peace: Inquiries into the History of the Cold War* (New York: Oxford University Press, 1989); John Mueller, The Essential Irrelevance of Nuclear Weapons: Stability in the Postwar World, *International Security* 13 (Autumn 1988): 55–79.

17. See Raymond L. Garthoff, *Detente and Confrontation: American-Soviet Relations from Nixon to Reagan,* rev. ed. (Washington, D.C.: Brookings, 1994), 228–42; Henry Kissinger, *White House Years* (Boston: Little, Brown, 1979), 183–94.

18. The best estimates are roughly 1,200 killed in the ten week conflict. The Indian army and air force suffered 474 killed, while regular Pakistani army casualties are estimated at 700 killed. Government of India, *Report of the Kargil Review Committee* (New Delhi: Government of India, 2000).

19. Van Evera, *Causes of War,* 245.

20. Barry R. Posen and Stephen Van Evera, "Defense Policy and the Reagan Administration: Departure from Containment," *International Security* 8 (Summer 1983): 33. According to Jervis, "The United States . . . should not have felt menaced by Soviet gains in the third world and should not have assumed that American security required the contraction of Soviet power." "Was the Cold War a Security Dilemma?" *Journal of Cold War Studies* 3 (Winter 2001): 54.

21. As early as the late 1960s, scholars had recognized the propensity of the superpowers to intervene in the third world. Responding to the thesis that bipolarity and nuclear weapons rendered these interventions unnecessary, Robert Osgood and Robert Tucker write, "[T]he fact remains that political realities do not conform to, and show no real prospect of conforming to, the results suggested by this analysis." *Force, Order, and Justice* (Baltimore: Johns Hopkins University Press, 1967), 279, cited in Robert Jervis, *System Effects: Complexity in Political and Social Life* (Princeton: Princeton University Press, 1997), 118n.

22. For a brief summary of some of these actions, see Garthoff, *Detente and Confrontation,* 732–45.

23. On the notion that the Soviet Union's assertiveness in the third world in the 1970s stemmed from its achievement of nuclear parity, see Bruce D. Porter, "Washington, Moscow, and Third World Conflict in the 1980s," in *The Strategic Imperative: New*

Policies for American Security, ed. Samuel P. Huntington (Cambridge, Mass.: Ballinger, 1982), 258–59; and Coit Blacker, "The Kremlin and Detente: Soviet Conceptions, Hopes, and Expectations," in *Managing U.S.-Soviet Rivalry,* ed. Alexander George (Boulder, Colo.: Westview, 1983), 127–28. For more on these Soviet actions in the third world, see Brian Crozier, *The Rise and Fall of the Soviet Empire* (Rocklin, Calif.: Forum, 1999), chaps. 21–39.

24. Garthoff, *Detente and Confrontation,* 744; Ellen C. Collier, *Instances of Use of United States Forces Abroad, 1798–1993* (Washington, D.C.: Congressional Research Service, 1993).

25. Jervis, *System Effects,* 120.

26. Van Evera, *Causes of War,* 178.

27. Moreover, as Robert Jervis notes, "The very fact that so little intrinsic value is at stake [in these interventions] means that each side's behavior reveals its general willingness to pay costs and run risks." Jervis, *System Effects,* 122.

28. McGeorge Bundy, "To Cap the Volcano," *Foreign Affairs* 48 (October 1969): 1–20.

29. Robert Jervis, "Why Nuclear Superiority Doesn't Matter," *Political Science Quarterly* 94 (Winter 1979–80): 617–33, at 618. Here Jervis does not share McGeorge Bundy's plea for superpower arms control negotiation: "It will take two to cap the volcano of strategic competition." Bundy, "To Cap the Volcano," 17.

30. Charles L. Glaser, "Realists as Optimists: Cooperation as Self-Help," *International Security* 19 (Winter 1994/95): 50–90, at 87 (emphasis added). See also Jervis, "Cooperation under the Security Dilemma," 188, 198; Van Evera, *Causes of War,* 244–45.

31. Jervis, "Cooperation under the Security Dilemma," 188.

32. For example, Donald MacKenzie argues that the U.S. development of highly accurate ICBMs was the result of a complicated amalgam of social and contextual forces. Donald MacKenzie, *Inventing Accuracy: A Historical Sociology of Nuclear Missile Guidance* (Cambridge, Mass.: MIT Press, 1990). MacKenzie's work is part of a wave of social constructivist literature on the determinants of technology; see Wiebe E. Bijker and others, *The Social Construction of Technological Systems: New Directions in the Sociology and History of Technology* (Cambridge, Mass.: MIT Press, 1999). The technological opportunism explanation advanced here shares the social constructivist rejection of strict technological determinism, but clearly places greater weight on the importance of relative material power than social context.

33. U.S. Department of State, *Memorandum by the Commanding General, Manhattan Engineer District (Groves),* January 2, 1946, in *Foreign Relations of the United States* [hereafter *FRUS*] (1946), 1:1203.

34. Ernest R. May, John D. Steinbruner, and Thomas W. Wolfe, *History of the Strategic Arms Competition, 1945–1972* (Washington: Department of Defense, 1981), 22.

35. U.S. Department of State, *Statement by President Truman at a Meeting at Blair House,* July 14, 1949, in *FRUS* (1949), 1:481–82.

36. See David Alan Rosenberg, "The Origins of Overkill: Nuclear Weapons and American Strategy, 1945–1960," *International Security* 7 (Spring 1983): 3–71; Marc Trachtenberg, "A 'Wasting Asset': American Strategy and the Shifting Nuclear Balance, 1949–1954," *International Security* 13 (Winter 1988–89): 5–48; Russell D. Buhite and Wm. Christopher Hamel, "War for Peace: The Question of an American Preventive War against the Soviet Union, 1945–1955," *Diplomatic History* 14 (Summer 1990): 367–84; Tami Davis Biddle, "Handling the Soviet Threat: 'Project Control' and the Debate on American Strategy in the Early Cold War Years," *Journal of Strategic Studies* 12 (Septem-

ber 1989): 273–302; George H. Quester, *Nuclear Monopoly* (New Brunswick, N.J.: Transaction, 2000).

37. Even traditional moderates in government, such as George Kennan, head of policy planning at the State Department, took seriously the question of whether it might be better for the United States to fight the Soviet Union during rather than after the era of American nuclear monopoly. Trachtenberg, "A 'Wasting Asset,'" 8–9. John Lewis Gaddis argues that Kennan saw preventive war only as a last resort, "to be considered if Soviet war-making potential was exceeding that of the United States and if opportunities for peaceful solutions had been exhausted." John Lewis Gaddis, *Strategies of Containment* (New York: Oxford University Press, 1982), 48–49n.

38. Trachtenberg, "Wasting Asset," 9–10.

39. Although the B-29 was the type of plane used to bomb Hiroshima and Nagasaki and government press releases at the time described the two squadrons as "atomic capable," these particular bombers had not as yet been modified to deliver atomic bombs. Scott D. Sagan, *Moving Targets: Nuclear Strategy and National Security* (Princeton: Princeton University Press, 1989), 15.

40. Walter Millis and E. Dufflied, eds., *The Forrestal Diaries* (New York: Viking Press, 1951), 455–58, 487; Buhite and Hamel, "War for Peace," 375; U.S. Department of State, *NSC-30: United States Policy on Atomic Warfare*, September 10, 1948, in *FRUS* (1948), 1:624–28.

41. Marc Trachtenberg, *A Constructed Peace: The Making of the European Settlement, 1945–1963* (Princeton: Princeton University Press, 1999), 89. In addition, the main purpose in instituting the blockade was not to bring political tensions to a head with a military crisis but to force the United States to reopen talks on the future of Germany and, specifically, to prevent the establishment of a West German state. Vladislav Zubok and Constantine Pleshakov, *Inside the Kremlin's Cold War: From Stalin to Khruschev* (Cambridge: Harvard University Press, 1996); Carolyn Eisenberg, *Drawing the Line: The American Decision to Divide Germany, 1944–1949* (Cambridge: Cambridge University Press, 1996).

42. Melvyn P. Leffler, *A Preponderance of Power: National Security, the Truman Administration, and the Cold War* (Stanford: Stanford University Press, 1992), 326. Also see Trachtenberg, *A Constructed Peace,* 89–90.

43. U.S. Department of State, *NSC-68: United States Objectives and Programs for National Security,* April 14, 1950, in *FRUS* (1950), 1:281–82.

44. Rosenberg, "Origins of Overkill," 25–26.

45. Trachtenberg, "Wasting Asset," 11–27. See also Dale C. Copeland, *The Origins of Major War* (Ithaca: Cornell University Press, 2000), 170–74.

46. Preventive war was still deemed preferable to the status quo national security policy. Nathan F. Twining, *Neither Liberty Nor Safety* (New York: Holt, Rinehart & Winston), 49; Trachtenberg, "Wasting Asset," 11–12.

47. Natural Resources Defense Council, *Archive of Nuclear Data, U.S. Nuclear Warheads, 1945–2002.* http://www.nrdc.org/nuclear/nudb/datainx.asp.

48. As the latter said to a reporter, "Give me the order to do it and I can break up Russia's five A-Bomb nests in a week . . . And when I went up to Christ I think I could explain to Him why I wanted to do it—now—before it is too late. I think I could explain to Him that I had saved civilization." Biddle, "Handling the Soviet Threat," 276–77.

49. Buhite and Hamel, "War for Peace," 376–77; Biddle, "Handling the Soviet Threat," 277.

50. "Minutes of the 71st meeting of the National Security Council," November 9, 1950, quoted in Trachtenberg, "Wasting Asset," 19.

51. Trachtenberg, "Wasting Asset," 20n.

52. Truman memorandum, January 27, 1952, President's Secretary's File (PSF), Papers of Harry S. Truman, Truman Library, quoted in Buhite and Hamel, "War for Peace," 378.

53. Buhite and Hamel, "War for Peace," 379. Although the Eisenhower administration conveyed to Chinese and Soviet leaders the American intention to escalate the conflict in Korea—and clearly this threat contributed to a breakthrough in the armistice talks that ended the war—scholars continue to disagree about the nature and impact of American atomic diplomacy during the Korean War. Compare, for example, Barry M. Blechman and Robert Powell, "What in the Name of God Is Strategic Superiority?" *Political Science Quarterly* 97 (Winter 1982–83): 589–602; Edward Keefer, "President Dwight D. Eisenhower and the End of the Korean War," *Diplomatic History* 10 (Summer 1986): 267–89; Richard K. Betts, *Nuclear Blackmail and Nuclear Balance* (Washington, D.C.: Brookings, 1987), 37–47; and Roger Dingman, "Atomic Diplomacy During the Korean War," *International Security* 13 (Winter 1988–89): 50–91.

54. U.S. Department of State, *Memorandum by the President to the Secretary of State,* September 8, 1953, in *FRUS* (1952–54), 2:461 (emphasis in original).

55. "General Matthew B. Ridgway, Memorandum for the Record," May 17, 1954, quoted in Rosenberg, "Origins of Overkill," 34.

56. U.S. Department of State, *Notes of NSC Meeting,* June 24, 1954, in *FRUS* (1952–54), 2:696.

57. For example, preventive war was considered following the Soviet hydrogen bomb test in 1953, after the French defeat in Vietnam in 1954, and during the Taiwan Straits crisis of 1954–55. See Buhite and Hamel, "War for Peace," 379–82; and Betts, *Nuclear Blackmail,* 48–62.

58. See Trachtenberg, "Wasting Asset," 32–46; Buhite and Hamel, "War for Peace," 382–84; and Copeland, *Origins of Major War,* 170–74.

59. U.S. Department of State, *Record of the 508th Meeting of the National Security Council,* January 22, 1963, in *FRUS* (1961–63), 8:462.

60. William Burr and Jeffrey T. Richelson, "Whether to Strangle the Baby in the Cradle": The United States and the Chinese Nuclear Program, 1960–64," *International Security* 25 (Winter 2000–2001): 54–99, and companion declassified documents posted at the National Security Archive. http://www.gwu.edu/nsarchiv/NSAEBB/NSAEBB38/; and Matthew Jones, " 'Groping Toward Coexistence': US China Policy during the Johnson Years," *Diplomacy and Statecraft* 12 (September 2001): 175–90.

61. *NSC-68,* 1:282.

62. David Alan Rosenberg, " 'A Smoking Radiating Ruin at the End of Two Hours': Documents on American Plans for Nuclear War with the Soviet Union, 1954–1955," *International Security* 6 (Winter 1981–82): 3–38; Rosenberg, "Origins of Overkill"; Sagan, *Moving Targets,* 22–24; and Scott D. Sagan, "SIOP-62: The Nuclear War Plan Briefing to President Kennedy," *International Security* 12 (Summer 1987): 22–51.

63. Rosenberg, " 'Smoking Radiating Ruin,' " 27.

64. Trachtenberg, *Constructed Peace,* 286.

65. Trachtenberg, *Constructed Peace,* chap. 8; David Rosenberg, "Reality and Responsibility: Power and Process in the Making of United States Nuclear Strategy, 1945–1968," *Journal of Strategic Studies* 9 (March 1986): 35–52. See also Francis J. Gavin, "The

Myth of Flexible Response: United States Strategy in Europe during the 1960s," *International Historical Review* 23 (December 2001): 847–75.

66. See, for example, William Burr, ed., "First Strike Options and the Berlin Crisis, September 1961: New Documents from the Kennedy Administration," in *National Security Archive Electronic Briefing Book No. 56*, National Security Archive, September 25, 2001. http://www.gwu.edu/nsarchiv/NSAEBB/NSAEBB56; Laurence Chang and Peter Kornbluh, eds., *The Cuban Missile Crisis, 1962: A National Security Archive Documents Reader* (New York: New Press, 1998); and William Burr, ed., "U.S. Planning for War in Europe, 1963–64," in *National Security Archive Electronic Briefing Book No. 31*, National Security Archive, May 24, 2000. http://www.gwu.edu/nsarchiv/NSAEBB/NSAEBB31/index.html.

67. Trachtenberg, *Constructed Peace*, 289–97; Fred Kaplan, "JFK's First-Strike Plan," *Atlantic Monthly*, October 2001, 81–86.

68. Trachtenberg, *Constructed Peace*, 292.

69. U.S. Department of State, *Memorandum of Conference with President Kennedy*, July 27, 1961, in *FRUS (1961–63)*, 8:123.

70. Carl Kaysen to General Maxwell Taylor, "Strategic Air Plan and Berlin," September 5, 1961, Record Group 218 (RG 218), Records of the Joint Chiefs of Staff, Records of Maxwell Taylor, National Archives; and Memorandum from Taylor to General Lemnitzer, September 19, 1961, RG 218, box 34, Records of Maxwell Taylor, Memorandums for the President, 1961. Both documents are available at Burr, *Electronic Briefing Book No. 56*.

71. Betts, *Nuclear Blackmail*, 104–9; Kaplan, "JFK's First-Strike Plan," 86.

72. Both Fred Kaplan and Marc Trachtenberg fall into this category of historians who have revised their views in the face of newly declassified documents. Kaplan, "JFK's First-Strike Plan"; Trachtenberg, *Constructed Peace*, 295n.

73. The following paragraph draws on Ernest R. May and Philip D. Zelikow, *The Kennedy Tapes: Inside the White House during the Cuban Missile Crisis* (Cambridge, Mass.: Belknap Press, 1997); Aleksandr Fursenko and Timothy Naftali, *'One Hell of a Gamble': Khrushchev, Castro, and Kennedy, 1958–1964* (New York: W.W. Norton, 1997); and Chang and Kornbluh, *The Cuban Missile Crisis*.

74. See Daryl G. Press, *Calculating Credibility: How Leaders Assess Military Threats* (Ithaca: Cornell University Press, 2005), chaps. 3–4.

75. Aleksandr G. Saveliyev and Nikolay N. Detinov, *The Big Five: Arms Control Decision-Making in the Soviet Union*, trans. Dimitryi Trenin, ed. Gregory Vorhall (Westport, Conn.: Praeger, 1995), 3.

76. U.S. Department of State, *Summary Record of the 517th Meeting of the National Security Council*, September 12, 1963, in *FRUS (1961–63)*, 8:499–507.

77. As John Mearsheimer describes the response of American policymakers to the emergence of MAD, "They were not only deeply unhappy about it, but for the remainder of the Cold War, they devoted considerable resources to escaping MAD and gaining a nuclear advantage over the Soviet Union." John J. Mearsheimer, *The Tragedy of Great Power Politics* (New York: W. W. Norton, 2001), 226. See also 128–33, 224–32.

78. In addition to sources previously cited, see Jervis, "Was the Cold War a Security Dilemma?" 36–60.

79. Robert S. Norris and Thomas B. Cochran, *US and USSR/Russian Strategic Offensive Nuclear Forces, 1945–1966* (Washington, D.C.: National Resources Defense Council, 1997). http://www.nrdc.org/nuclear/nudb/datainx.asp.

80. In 1959, acknowledging that the United States might still suffer unimaginable destruction after a massive first strike, Eisenhower concluded, "All we really have that is meaningful is a deterrent." Quoted in Rosenberg, "The Origins of Overkill," 62–63.

81. Most analysts cite as a baseline Defense Secretary Robert McNamara's famous criteria for an assured destruction capability developed in 1963. McNamara's judgment (which was accepted by Presidents Kennedy and Johnson and the Congress) was that, after a surprise Soviet first strike, the United States needed enough surviving forces to destroy roughly 30 percent of the Soviet population and 70 percent of their industrial capacity, which entailed destroying 150 of their largest cities. McNamara and his aides determined that this task could be achieved by delivering around four hundred one-megaton warheads. Anything beyond four hundred warheads would not significantly increase the level of damage inflicted on the Soviet Union, reflecting diminishing marginal returns. To be very conservative, assuming the need for at least four hundred survivable warheads on each leg of the triad, the United States would need about fifteen hundred warheads in its arsenal for assured destruction capability. For a discussion of McNamara's criteria, see Alain C. Enthoven and K. Wayne Smith, *How Much Is Enough? Shaping the Defense Program, 1961–1969* (New York: Harper & Row, 1971), 175, 207–10; Fred Kaplan, *The Wizards of Armageddon* (Stanford: Stanford University Press, 1983), chap. 22; and Sagan, *Moving Targets*, 32–34.

82. Norris and Cochran, *US and USSR/Russian Strategic Offensive Nuclear Forces;* and John Pike, "Nuclear Forces Guide," Federation of American Scientists. http://www.fas .org/nuke/guide/summary.htm.

83. See Michael Salman, Kevin J. Sullivan, and Stephen Van Evera, "Analysis or Propaganda? Measuring American Strategic Nuclear Capability, 1969–88," in *Nuclear Arguments: Understanding the Strategic Nuclear Arms and Arms Control Debates,* ed. Lynn Eden and Steven E. Miller (Ithaca: Cornell University Press, 1989), 172–263, at 210–11.

84. The Soviet Union deployed 2,327 warheads in 1970; 7,488 in 1980; and 11,252 in 1990. The growth of the superpower arsenals is even more staggering if one considers all stockpiled nuclear warheads. For the United States, these totals were 369 in 1950; 20,434 in 1960; 25,742 in 1970; 27,958 in 1973 (the height); 23,387 in 1980; and 20,684 in 1990. For the Soviet Union, total stockpiled warheads were 5 in 1950; 1,605 in 1960; 11,643 in 1970; 30,062 in 1980; 40,723 in 1986 (the height); and 33,515 in 1990. Norris and Cochran, *US and USSR/Russian Strategic Offensive Nuclear Forces.*

85. See Aaron L. Friedberg, "The Evolution of U.S. Strategic Doctrine, 1945–1980," in Huntington, *The Strategic Imperative,* 53–99; Desmond Ball, "The Development of the SIOP, 1960–1983," in *Strategic Nuclear Targeting,* ed. Desmond Ball and Jeffrey Richelson (Ithaca: Cornell University Press, 1986), 57–83; Sagan, *Moving Targets;* Eric Mlyn, *The State, Society, and Limited Nuclear War* (Albany, N.Y.: SUNY Press, 1995); and Lawrence Freedman, *The Evolution of Nuclear Strategy,* 3rd ed. (Houndmills, U.K.: Palgrave, 2003).

86. Rosenberg, "The Origins of Overkill"; Friedberg, "Evolution of U.S. Strategic Doctrine"; and Jervis, *Illogic of American Nuclear Strategy,* 44.

87. Ball, "Development of the SIOP," 66–67.

88. A study for Defense Secretary McNamara in 1964 demonstrated that for each extra dollar that the Soviets added to the attack forces, the United States would have to spend $3.00 to protect 70 percent of its industry, $2.00 to save 60 percent, $1.80 to defend 50 percent, and $1.00 to defend 40 percent. In short, "in the race between offense and defense, offense would win, and at lower cost." Kaplan, *Wizards of Armageddon,* 321–22.

89. During the late 1960s and 1970s, many argued that the United States had abandoned counterforce targeting and accepted the military and political implications of MAD. However, as John Mearsheimer writes, "It is now well established among students of the nuclear arms race that this claim is a groundless myth perpetrated by experts and policymakers who surely knew better." *Tragedy of Great Power Politics,* 484n.

90. For more on these systems, see Miller, *Cold War,* chaps. 9–10.

91. Posen and Van Evera, "Defense Policy and the Reagan Administration, 3–45; and Desmond Ball and Robert C. Toth, "Revising the SIOP: Taking War-Fighting to Dangerous Extremes," *International Security* 14 (Spring 1990): 65–92.

92. See Robert P. Berman and John C. Baker, *Soviet Strategic Forces: Requirements and Responses* (Washington, D.C.: Brookings, 1982), chap. 3; David Holloway, *The Soviet Union and the Arms Race* (New Haven: Yale University Press, 1983), chap. 3; William T. Lee, "Soviet Nuclear Targeting Strategy," in Ball and Richelson, *Strategic Nuclear Targeting;* and Miller, *Cold War,* 98–102.

93. Lee, "Soviet Nuclear Targeting Strategy," 85; Miller, *Cold War,* 101.

94. The phrase is from Bundy, "To Cap the Volcano," 20.

95. Some of these arguments are stated in Glaser, "When Are Arms Races Dangerous?" 74–81.

96. Among a series of influential articles by Paul Nitze in the mid-1970s, which attempted to specify criteria for effective deterrence but which perplexed advocates of MAD, see Paul H. Nitze, "Assuring Strategic Stability in an Era of Détente," *Foreign Affairs* 54 (January 1976): 207–32. "Throw weight" is a term for the useful weight that can be placed on a trajectory toward the target by the main or boost stage of a missile, which would include the total weight of reentry vehicles, post-boost-stage vehicles, and penetration aids. Miller, *Cold War,* 452.

97. Ted Greenwood, *Making the MIRV: A Study of Defense Decision Making* (Cambridge, Mass.: Ballinger, 1975). For arguments that MIRV was not necessary for these purposes, see Jervis, *Illogic of American Nuclear Strategy;* and Glaser, *Analyzing Strategic Nuclear Policy,* chap. 7. Charles Glaser discusses this and other U.S. cold war arms buildup decisions in Glaser, "When Are Arms Races Dangerous?" 74–81.

98. Daniel Buchonnet, "MIRV: A Brief History of Minuteman and Multiple Reentry Vehicles," Lawrence Livermore Laboratory, February 1976, 9, 12, National Security Archive, *U.S. Nuclear History: Nuclear Arms and Politics in the Missile Age, 1955–1968.* http://www.gwu.edu/nsarchiv/nsa/NC/mirv/mirv1_1.html (emphasis added).

99. Mearsheimer, *Tragedy of Great Power Politics,* 231.

100. The entire "security-seeking" versus "greedy state" distinction commonly used in realist scholarship is misleading because the logic of both technological opportunism and offense-defense theory suggests that all states are seeking to maximize their security. The two schools simply differ on how states pursue that objective.

101. Earl C. Ravenal, "Counterforce and Alliance: The Ultimate Connection," *International Security* 6 (Spring 1982): 26–43.

102. The chief U.S. negotiator at the 1971 SALT negotiations urged the White House to support the emerging ABM ban because "it would be militarily advantageous to the U.S.," given that U.S. Poseidon SLBMs would have an easier time striking key Soviet targets. However, given Poseidon capabilities (relatively small and inaccurate reentry vehicles) and the fact that Gerard Smith was a prominent MAD proponent, it is likely that this comment concerns the Poseidon role as a second-strike system. Message

from Gerard C. Smith to Henry A. Kissinger, August 7, 1971, archived in William Burr, ed., "Missile Defense Thirty Years Ago: Déjà Vu All Over Again?" in *National Security Archive Electronic Briefing Book, No. 36*, National Security Archive, December 18, 2000. http://www.gwu.edu/nsarchiv/NSAEBB/NSAEBB36.

103. See David Goldfischer, *The Best Defense: Policy Alternatives for U.S. Nuclear Security from the 1950s to the 1990s* (Ithaca: Cornell University Press, 1993); and Frances FitzGerald, *Way Out There in the Blue: Reagan, Star Wars, and the End of the Cold War* (New York: Simon and Schuster, 2000).

104. Quoted in Fareed Zakaria, "Misapprehensions about Missile Defense," *Washington Post*, May 7, 2001.

Conclusion

1. Richard K. Betts, "The Soft Underbelly of American Primacy: Tactical Advantages of Terror," *Political Science Quarterly* 117 (Spring 2002): 19–36, at 20.

2. Ibid., 33.

3. Carl von Clausewitz, *On War*, ed. and trans. Michael Howard and Peter Paret (Princeton: Princeton University Press, 1976), 358.

4. Kenneth Waltz argues that the drive for competition in a self-help international system means that successful weapons and military practices will be imitated quickly. "The fate of each state depends on its responses to what other states do. The possibility that conflict will be conducted by force leads to competition in the arts and the instruments of forces. Competition produces a tendency toward the sameness of the competitors. . . . Contending states imitate the military innovations contrived by the country of greatest capability and ingenuity. And so the weapons of major contenders, and even their strategies, begin to look much the same all over the world." Kenneth N. Waltz, *Theory of International Politics* (New York: McGraw-Hill, 1979), 127. For an argument about the systemic consequences of revolutionary military capabilities and practices, see Emily O. Goldman and Richard B. Andres, "Systemic Effects of Military Innovation and Diffusion," *Security Studies* 8 (Summer 1999): 79–125.

5. See Martin van Creveld, *Technology and War: From 2000 B.C. to the Present* (New York: Free Press, 1991).

6. See David E. Johnson, *Fast Tanks and Heavy Bombers: Innovation in the U.S. Army, 1917–1945* (Ithaca: Cornell University Press, 1998).

7. Sean M. Lynn-Jones, "Offense-Defense Theory and Its Critics," *Security Studies* 4 (Summer 1995): 660–91, at 667.

8. George H. Quester, *Offense and Defense in the International System* (New York: John Wiley & Sons, 1977), 10–11; Lynn-Jones, "Offense-Defense Theory and Its Critics," 686–87; Charles L. Glaser and Chaim Kaufmann, "What Is the Offense-Defense Balance and Can We Measure It?" *International Security* 22 (Spring 1998): 44–82, at 48–49; Stephen Van Evera, *Causes of War: Power and the Roots of Conflict* (Ithaca: Cornell University Press, 1999), 4–5.

9. See Jack Snyder, *The Ideology of the Offensive: Military Decision Making and the Disasters of 1914* (Ithaca: Cornell University Press, 1984); and Barry R. Posen, *The Sources of Military Doctrine: France, Britain, and Germany between the World Wars* (Ithaca: Cornell University Press, 1984).

10. See John J. Mearsheimer, *The Tragedy of Great Power Politics* (New York: W. W. Norton, 2001), chap. 5.

11. Bernard Brodie, "Technological Change, Strategic Doctrine, and Political Outcomes," in *Historical Dimensions of National Security Problems,* ed. Klaus Knorr (Lawrence: University Press of Kansas, 1976), 263–306 at 263.

Index

Acheson, Dean, 6
Afghanistan War (2001), 15
aircraft carriers, 36
alliance behavior, 2, 30–32, 182 n5, 186 n26, 187 n31
Alsace-Lorraine, 74, 93
American Civil War (1861–65): railroads and, 47, 49–55, 60, 64–65, 67, 78; small arms/artillery and, 81–82, 85–86, 90
Amiens offensive (1918), 101
"Anaconda Plan," 52
Anderson, Orvil, 136
Antiballistic Missile (ABM) Treaty, 143, 146
Antietam, battle of (1862), 82
Ardennes battles (1943–44), 104–5, 110, 114
armored warfare, 23, 99–122; blitzkrieg and, 32, 99–100, 115–19, 157, 210 n76; "deep battle" and, 108, 110; distinguishability of offensive/defensive weapons in, 36–37; flanking/enveloping operations and, 109–10; in Italo-Ethiopian War, 102; military outcomes and, 23, 100–114, 121, 151; mobility-favors-offense hypothesis and, 99, 154; in Moroccan crises, 102; political outcomes and, 23, 100, 114–22; in Russo-Japanese border clashes, 102; in Spanish Civil War, 102; in World War I, 100, 115, 117, 121; in World War II, 23, 32, 100–101, 103–14, 121

arms control, 10–12, 13–15, 35–37, 132–33, 150, 155
arms racing, 7, 9–10, 23, 124–25, 132–48, 157. *See also* security dilemma
artillery. *See* small arms and artillery
Atlanta campaign (1864), 52–53, 85
atomic bombs. *See* nuclear weapons
Austria. *See specific wars/conflicts*
Austro-Prussian War (1866), 23, 52, 55–57, 69–71, 82–84, 91

Balaklava, battle of (1854), 81
Baruch Plan, 134
battles. *See specific battles* (*e.g.*, Antietam, battle of)
Belgium, 60–61, 96, 104, 119–20
Belorussia (1943), 113
Benedek, Ludwig, 56
Berlin blockade (1948–49), 135
Berlin crisis (1961–62), 138–39
Bethmann Hollweg, Theobald von, 76, 97
Betts, Richard, 15–16, 150
Biddle, Stephen, 17, 87, 182 n2, 191 n8
Bismarck, Otto von, 65, 68–71, 74, 91, 93, 157
blitzkrieg, 32, 99–100, 115–19, 157, 210 n76
Bock, Fedor von, 120
Boer War. *See* British-Boer War
Boggs, Marion William, 7, 36–37, 42
bombers, 4, 36

Cornell Studies in Security Affairs

A series edited by ROBERT J. ART
ROBERT JERVIS
STEPHEN M. WALT

The Wrong War: American Policy and the Dimensions of the Korean Conflict, 1950–1953
  by Rosemary Foot
The Best Defense: Policy Alternatives for U.S. Nuclear Security from the 1950s to the 1990s
  by David Goldfischer
Storm of Steel: The Development of Armor Doctrine in Germany and the Soviet Union, 1919–1939
  by Mary R. Habeck
America Unrivaled: The Future of the Balance of Power
  Edited by G. John Ikenberry
The Meaning of the Nuclear Revolution: Statecraft and the Prospect of Armageddon
  by Robert Jervis
Fast Tanks and Heavy Bombers: Innovation in the U.S. Army, 1917–1945
  by David E. Johnson
Modern Hatreds: The Symbolic Politics of Ethnic War
  by Stuart J. Kaufman
The Vulnerability of Empire
  by Charles A. Kupchan
The Transformation of American Air Power
  by Benjamin S. Lambeth
Anatomy of Mistrust: U.S.–Soviet Relations during the Cold War
  by Deborah Welch Larson
Planning the Unthinkable: How New Powers Will Use Nuclear, Biological, and Chemical Weapons
  edited by Peter R. Lavoy, Scott D. Sagan, and James J. Wirtz
Cooperation under Fire: Anglo-German Restraint during World War II
  by Jeffrey W. Legro
Dangerous Sanctuaries: Refugee Camps, Civil War, and the Dilemmas of Humanitarian Aid
  by Sarah Kenyon Lischer
Uncovering Ways of War: U.S. Intelligence and Foreign Military Innovation, 1918–1941
  by Thomas Mahnken
No Exit: America and the German Problem, 1943–1954
  by James McAllister
Liddell Hart and the Weight of History
  by John J. Mearsheimer
Reputation and International Politics
  by Jonathan Mercer
Undermining the Kremlin: America's Strategy to Subvert the Soviet Bloc, 1947–1956
  by Gregory Mitrovich
The Remnants of War
  by John Mueller
Report to JFK: The Skybolt Crisis in Perspective
  by Richard E. Neustadt
The Sacred Cause: Civil-Military Conflict over Soviet National Security, 1917–1992
  by Thomas M. Nichols